Biology of the Plant Rusts

AN INTRODUCTION

Biology of the Plant Rusts

AN INTRODUCTION

LARRY J. LITTLEFIELD

Iowa State University Press / Ames

Larry Littlefield, Program Specialist in International Training, U.S. Department of Agriculture, spent 15 years on the staff of North Dakota State University, Department of Plant Pathology. There he taught mycology and plant pathology. His research focused on anatomy and ultrastructure of rust fungi and host-parasite relations, culminating in the book *Ultrastructure of Rust Fungi,* coauthored by M. C. Heath, published in 1979. Dr. Littlefield received the Ph.D. degree from the University of Minnesota. He also spent a year each at the University of Uppsala, Purdue University, and Oxford University in postdoctoral and sabbatical research.

Composed and printed by The Iowa State University Press, Ames, Iowa 50010

First edition, 1981

Library of Congress Cataloging in Publication Data

Littlefield, Larry J.
Biology of plant rusts.

Includes indexes.
1. Rusts (Fungi) 2. Fungous diseases of plants.
I. Title.
SB741.R8L39 632′.425 81-3734

ISBN 0-8138-1670-X AACR2

This book is dedicated, with thanks and respect, to my teacher of undergraduate plant pathology and first professional mentor

---

DR. CARL W. BOOTHROYD

CONTENTS

PREFACE

THE RUST FUNGI and rust diseases constitute a significant portion of many university courses in mycology and plant pathology, and to a lesser extent, courses in botany and microbiology. Hundreds of research papers have been written about the rusts. Several books have covered such specific topics as taxonomy, identification, ultrastructure, spore germination, control, and host-parasite relations, either in individual volumes or as chapters in multiauthored volumes. *The Plant Rusts* (*Uredinales*), by J. C. Arthur (John Wiley & Sons, 1929), is the most recent book to cover the spectrum of rust biology as it was then known.

The undergraduate student cannot possibly sift through the vast material available on the rusts. It is for those students that this book is intended. I have tried to survey the whole field of rust biology, to sort out the essential information, and to organize it in a brief, readable manner. Obviously, much information had to be excluded. In addition to the References are supplemental reading lists to help guide those students who desire more information. These readings are often review papers or other books that were selected, in part, for their large number of literature citations as well as their summary nature. It is my hope that this book will spark the interest of beginning students and stimulate them to study further this fascinating group of fungi.

I thank numerous colleagues for reading portions of the manuscript and making many helpful suggestions. They include G. B. Cummins, M. C. Heath, D. H. Lewis, A. P. Roelfs, R. Rohringer, and especially J. A. Browning and J. D. Miller. Any errors or misinterpretations that occur are wholly my responsibility.

I also thank M. F. Brown, R. E. Gold, R. Guggenheim, J. Harr, Y. Hiratsuka, V. A. Johnson, K. Mendgen, J. D. Miller, T. Nicholls, H. R. Powers, A. P. Roelfs, D. M. Spencer, G. D. Statler, F. H. Tainter, J. R. Venette, and J. Walker for their photographs; Matt Lanz for assistance with the drawings; and Jan Hogfoss, Nancy Brager, and Sharon Gregor for typing the manuscript.

Biology of the Plant Rusts

AN INTRODUCTION

1 Introduction

1.1 Host Range; Geographic Range

The rust fungi, numbering some 5000 species in about 130 genera, attack an extremely wide range of hosts, including ferns, Gymnosperms, and mono- and dicotyledonous Angiosperms. Rusts have not been reported on mosses, liverworts, algae, horsetails, or other primitive plants such as *Lycopodium* spp. and *Psilotum* spp. The fossil record shows that rust or rustlike fungi date back to the Cretaceous period, ca. 130–150 million years, and possibly to the Pennsylvanian period, ca. 280–325 million years.

Geographically, rusts occur worldwide, although some are restricted climatically. *Ravenelia, Diabole, Phakopsora, Prospodium,* and numerous other genera are restricted to warmer climates. Other genera, e.g., *Pucciniastrum* and *Uredinopsis,* occur more often in alpine to boreal habitats. The genus *Puccinia* occurs over a broad climatic range. The climatic-geographic restriction of rusts depends somewhat on similar restrictions on their hosts. Exceptions to this occur where certain rusts appear on different host species in totally different climatic regions. For example, *Coleosporium asterum* is present in its uredinial and telial stages on goldenrod far south of where its pycnial and aecial stages occur on pine.

1.2 Economically Important Rusts

Economically, rusts have been a major factor in agricultural and silvacultural productivity for centuries. *Cronartium ribicola* (white pine blister rust) (see Plate 2.4), a rust indigenous to Siberia, spread into Europe in the late 1800s and decimated forests of 5-needle pine, including the Weymouth pine in Great Britain. This pine, prized for lumber and ship masts, had previously been introduced into Europe from North America, where it is known as eastern white pine. Soon after the rust spread to Europe from Asia it was introduced on nursery stock into North America, where it eventually extended across that continent. Quarantine laws enacted by the United States in 1912 were too late to prevent the millions of dollars lost to this rust continuing to the present time.

Other rusts also continue to cause millions of dollars of loss annually. Notable examples are *Cronartium coleosporioides* (stem and branch cankers) on 2- and 3-needle pines, *C. fusiforme* (fusiform stem canker) (Fig. 1.1) in numerous species of young pines in southern United States, *Endocronartium harknessii* (western gall rust) (Fig. 1.2) on various pines in western North America, and *Melampsora pinitorqua* (stem twisting rust) on various pines throughout Europe. Fusiform rust was estimated to reduce the value of the 1972 loblolly and slash pine harvest in southeastern United States by about $28 million (Fig. 1.1). Somewhat less severe economically, but important for aesthetic reasons, is the rust

Fig. 1.1. Extensive damage in loblolly pine plantation from fusiform rust infection, *Cronartium fusiforme*. (Courtesy H. R. Powers et al. by permission of the Society of American Foresters)

Fig. 1.2. Western gall rust, *Endocronartium harknessii,* has killed many branches in this mature lodgepole pine. (From Ziller 1974, reproduced by permission of the Minister of Supply and Services Canada)

on Australian acacia (wattle) caused by *Uromycladium* spp. In addition to forming galls and causing unsightly disfigurations in ornamental plantings (see Fig. 4.9), the rust can kill trees and ruin entire plantations. *Melampsora larici-populina* causes a leaf rust on various poplar species in Europe and elsewhere. That rust had a significant indirect effect on Argentina's fruit industry early this century. There, plantations of introduced poplar supplied wood for overseas containers used for fruit export, and for many years lumber production for these crates was markedly reduced by poplar rust until resistant clones of poplar were introduced. Poplar rust remains an important disease in Australia, having first appeared there in 1972.

Rusts of wheat, maize, barley, oats, and other cereals around the world are extremely costly diseases. Fortunately, rust is not a serious problem on rice, although two species are pathogenic on that plant.

In Europe the most significant and widespread rust pathogen of wheat is *Puccinia striiformis,* which causes the disease known as yellow rust. In England losses have been estimated to be 5–10% of the total crop in an epidemic year and up to 50–60% reduction in susceptible varieties under conditions ideal for the disease. In The Netherlands 70% of the 1956 winter wheat crop was destroyed by yellow rust. In North America this disease is known as stripe rust. It is restricted essentially to the western provinces of Canada, along the Pacific coast and northwestern intermountain regions of the United States, and southward into the mountains of Mexico and Guatemala. In the United States, losses in commercial fields in 1960 and 1961 often ranged from 20 to 75%, with losses of some 25 million bushels (680,000 metric tons) in 1961.

More important to North American wheat production is *Puccinia graminis* f. sp. *tritici,* the so-called black stem rust fungus (Figs. 1.3, 1.4; see also Fig. 2.17, Plate 2.6). Sporadic epidemics of stem rust have cost North American farmers countless billions of dollars. In 1916 the disease was estimated to have destroyed 300 million bushels (8.2 million metric tons) of wheat in the United States and Canada. The 1935 epidemic destroyed at least 135 million bushels (3.7 million metric tons), mostly in the spring wheat areas of the Dakotas and Minnesota. During the decade of the 1950s the average loss from stem rust in the United States was estimated to be about 4%, or 40 million bushels (1.1 million metric tons), annually. In a state such as North Dakota, where wheat is the single largest source of income, the extreme economic impact of such epidemics is immediately apparent. Although serious epidemics of stem rust in the winter wheats of Kansas and Nebraska caused losses of 35 and 32 million bushels (about 0.9 million metric tons) in 1961 and 1962 respectively, there has been no major epidemic in the spring wheat regions of North America since the early 1950s, thanks in part to vigilant efforts of plant breeders and pathologists. Partly as a result of this success, the public often takes rust-resistant cultivars for granted—a potentially dangerous attitude.

The cost for continued breeding programs may seem high. Nevertheless, recent figures show that the estimated $217 million annual value of wheat stem rust resistance in western Canada provides, conservatively, about an 8:1 benefit:cost ratio. Similar benefits of continued research can be cited in Australia, where no major losses in hard wheat in northern New South Wales have occurred for some 30 years. In contrast, little has been done until recently on rust control in other parts of Australia where climatic conditions occasionally permit major epidemics of rust. In 1973 the loss was estimated at $200 million in southern New South Wales.

A classic disease that caused ruination of countless plantation owners, bankrupted the Oriental Bank, and changed drinking habits throughout the former British Empire is coffee rust, caused by *Hemileia vastatrix.* Until the late nineteenth century, coffee was the major source of caffeine in the empire, being grown in Ceylon (Sri Lanka) and other colonies in southern Asia. The widespread, rapidly advancing epidemic of coffee rust in the 1870s and 1880s devastated coffee plantations. They were replaced with tea plantations, and tea replaced coffee as a source of caffeine for millions of English and other people of the empire.

Fig. 1.3. Devastation of wheat caused by *Puccinia graminis* f. sp. *tritici.* Heavy infection of the cultivar in the center row results in death of stems and leaves. Healthy, rust-resistant cultivars are on either side. (Courtesy J. D. Miller, U.S. Department of Agriculture)

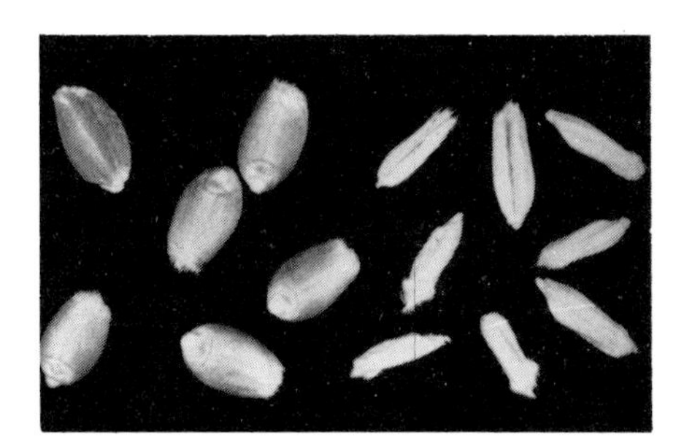

Fig. 1.4. Extensive shriveling of wheat seed (right) caused by *Puccinia graminis* f. sp. *tritici* infection of stems and leaves compared to seed from uninfected plant (left). (Courtesy J. D. Miller, U.S. Department of Agriculture)

The coffee rust fungus later spread to African plantations, where it has been estimated to cause annual losses of some 4.5 million kg of coffee beans in Kenya. It continues to threaten world coffee production following its introduction in the 1960s into South America from Africa, presumably by transatlantic winds. Projected potential losses from coffee rust in Central America, Panama, and Mexico have been calculated to range from $22 to $132 million annually, based on 5 and 30% yield loss respectively. Coffee rust epidemics in Latin America could have profound effects upon the national economies of those nations; e.g., the percentage of foreign exchange provided by coffee sales in Colombia, El Salvador, and Brazil are 68, 43, and 41% respectively.

Soybean rust, caused by *Phakopsora pachyrhizi,* has been described as the greatest threat to soybean production in eastern Australia. It causes some 20–30% annual loss to the crop in Taiwan, with up to 70–80% losses in some instances. As yet, this major pathogen has not appeared in continental North America, although it does occur in Puerto Rico.

The arrival of sugarcane rust (*Puccinia melanocephala*) in North America in 1979 has placed a new burden on American sugar production.

Another potentially serious rust, caused by *Puccinia pittieriana,* occurs on potato. It can cause significant yield losses in certain higher elevations of the Andes in South America. The cool climate required for potato rust development is similar to that of many areas of Ireland and the United Kingdom. Since *P. pittieriana* in South America is a problem only at elevations 2700–3400 m above sea level, its potential as a pathogen in Europe or North America is unknown.

Other serious or potentially serious rusts occur on bean, maize, flax, sunflower, and cotton, to mention a few. Nor is the home gardener or nursery grower free from rust problems, e.g., blackberry, chrysanthemum, hollyhock, apple, and rose rusts.

1.3 History of Plant Rusts

Recognition of rusts as such by the ancients is questionable. Although several claims state that rusts and blasts are mentioned in the Old Testament, the general decline and/or pestilence of crops referred to may have resulted from a number of causes. For example, "blasting" may have been the product of hot sirocco winds off the Arabian desert rather than of rust fungi. A major difficulty in interpreting such ancient writings is the lack of precise equivalents in modern languages; e.g., the Hebrew word "yērākôn" has been translated variously to mildew, jaundice, or paleness. Also, as late as the nineteenth century in Great Britain, the term "mildew" was commonly used for what is now called "rust."

The observations and descriptions by Aristotle (384–322 B.C.) were more precise than those of his Hebrew predecessors. For example, he noted that rust (as well as can be equated) was more severe in some years than in others, but he attributed it to warm vapors. Aristotle's student, Theophrastus (371–286 B.C.), was more exacting yet. He recorded the particular susceptibility of cereals to rust compared to legumes and that barley, especially of the Achillean type, was the most susceptible. He noted that elevated fields or those exposed to wind were less likely to rust than low-lying fields.

The most lucid accounts of rust and the associated religious ceremonies were provided by writers of the early Christian era. According to Ovid (43 B.C.–17 A.D.), legend said that rust was visited upon mankind as punishment for the burning of a live fox (caught in a chicken yard!); it was wrapped in straw and turned loose by a vengeful farm youth. Unfortunately for countless successors of that animal, the tradition of igniting torches tied to the tails of foxes set loose in the Circus of Rome became established as an annual ritual on April 19 as a means of averting rust. However, a more sacred festival, the Robigalia, was practiced annually from about 700 B.C. This ceremony, founded by Numa Pompilius, second king of the Romans, was centered around the rust god, variously named Robigo (female) or Robigus (male). The annual date of the Robigalia was April 25, when the wheat began to head. This most solemn ceremony is described by Ovid, who happened by chance to witness it in the holy grove of the rust god Robigus, some 5 miles from Rome near the Claudian Way. There, amid prayers and supplications by white-robed priests, libations of wine and incense and the flaming sacrifice of the entrails of a red dog (symbolic of the red rust) and a sheep were placed upon the alter. The pagan ceremony of the Robigalia was later incorporated into Christian tradition as the annual rite of St. Mark's Day, or Rogation, and is still observed on April 25.

Distribution of the wheat stem rust fungus to northern Europe did not occur for several centuries. Presumably during Moslem expansions of the eighth and ninth centuries, the barberry plant, the alternate host of the stem rust fungus, was introduced into northern Europe. Whether the rust fungus came with the barberry or became established on it from airborne inoculum is not known.

It was not until several centuries after that, that the association between rust on cereals and the small yellow blisters on the barberry leaves was proven. However, astute observations by French farmers that cereal rust was more severe near barberry bushes led to the first legislation related to the control of plant diseases. In Rouen, France, in 1660, a law was passed requiring the destruction of barberry plants near grain fields. That was followed by similar laws in the English colonies of North America (the barberry having been brought to the New World) and much of Europe. The hardy barberry survived, however. Not only did it continue to plague the farmers as a source of the rust fungus, it presented a formidable challenge to plant pathologists of the nineteenth and early twentieth centuries, who attempted and ultimately succeeded in elucidating the entire life cycle of the fungus. The final page in the complete description of the life cycle of the wheat rust fungus (see Chap. 2) was not written until 1927 in Winnipeg, Canada, some two millenia after the lucid, but rather bloody, description of the Robigalia on the outskirts of Rome.

1.4 Human Contributions to Rust Epidemics

The most obvious contribution to rust epidemics is the introduction of rusts by people from one region of the world into another. This is particularly dangerous, since in the new environment there has not been the opportunity for coevolution of natural biological controls of the rust, nor in the host population has there been natural selection for resistance. Consequently, the potential exists for rampant spread of the pathogen in epidemic proportions. Already mentioned is white pine blister rust, which was introduced into North America from Europe on diseased nursery stock. Quite recently, white rust of chrysanthemum (*Puccinia horiana*) was distributed from Japan to Australia and the United States. Introduction of this rust to Europe was via South Africa on diseased cuttings from one major supplier. In a recent U.S. Department of Agriculture compilation of some 1500 foreign potential bacterial and fungal diseases of food, fiber, and forage crops, more than 400 rusts were included. Vigilence must be maintained to prevent repetition of past mistakes.

Economic demands of modern intensive agriculture encourage use of genetically uniform cultivars in large fields over large geographic areas (Fig. 1.5). For windborne pathogens such as rusts, this greatly increases the chances of widespread epidemics, i.e., pandemics, often spanning several countries. Most notable examples of such epidemics are those of black stem rust of wheat in North America; yellow rust of wheat in Europe; and coffee rust in Asia, Africa, and South America. In the case of North American hard red spring and winter wheats, together totaling some 50 million acres (20.25 million hectares), often less than half a dozen cultivars constitute the vast majority of acreage. The dangers inherent in this practice are obvious, and alternatives to widespread genetic uniformity are imperative. In the developing nations of Africa, Asia, and South America, most of the new, highly productive semidwarf cultivars of wheat originated from the breeding program of CIMMYT (International Maize and Wheat Improvement Center, Mexico City). This is not to say, however, that all those cultivars are dangerously homogeneous. CIMMYT is making diverse genetic materials available rapidly, as

Fig. 1.5. Wheat harvest in the Great Plains region of the United States. Vast areas of genetically uniform cultivars present major problems in plant disease control. (Courtesy V. A. Johnson, U.S. Department of Agriculture)

evidenced by the release of two CIMMYT-type multiline varieties (see Chap. 7.5) of wheat in India in 1978.

Extreme genetic homogeneity exists in the background of coffee in Central and South America. There, all the original plantations were derived from one bush in the Amsterdam Botanic Garden or from a seedling of that bush sent to the Botanic Garden in Paris. Although coffee-breeding programs in Brazil have produced new cultivars containing germ plasm from native African lines, most cultivars used are genetically related and differ by only a few genes.

Our disturbance of natural ecosystems has increased the destructiveness of certain forest rusts without the introduction of any new host or pathogen germ plasm. Fusiform rust (*Cronartium fusiforme*) of slash and loblolly pine in southeastern United States (Fig. 1.1) was a rarity in natural stands before 1900. It is now epidemic in pine plantations throughout that region, having increased dramatically over the past 30 years. This increase has been due to management practices that, among other things, have resulted in: (1) replacement of large areas of more resistant longleaf pine with faster growing but more susceptible slash and loblolly pines; (2) the presence of more succulent and susceptible tissue because of intensive site preparation, irrigation, and fertilization; genetic selection for rapid growth; reduced age diversity and average age of trees, now 8–10 years compared to 35 years in 1900; (3) extending the range of slash and loblolly pines into areas where oaks, the alternate host, grow; and (4) the establishment and growth of oaks by intensive fire control practices.

Similarly, human disturbance of genetic systems has brought to the fore rust problems that did not previously exist. Stripe rust was not a major problem in northwestern United States until 1961, when the older varieties of wheat had been replaced by newer ones. During the development of these new varieties, many of the genes for resistance were lost; consequently, *Puccinia striiformis* became economically important although it had undergone no evident genetic change.

Despite the monumental strides made in breeding for rust resistance, only recently have breeders and pathologists realized that much of their earlier work inadvertently "set up" many new cultivars for rapid devastation by new races of rust pathogens. For decades, selection for resistance was based, understandably, on highly specific, clearly recognized immunity or a high degree of resistance, which is often controlled by a single gene. Unfortunately, genetic variability of the pathogen frequently gave rise to new races capable of attacking the new cultivars. This repeated development of pathogenic races in response to the release of successive new varieties has been aptly termed "man-guided evolution." Selection for resistance in breeding programs is currently shifting toward genetically broader, more generalized types of resistance, away from the previously prized specific immunity (see Chap. 7.5). It is hoped that cultivars developed by this procedure will be longer lived than were many of their predecessors.

1.5 Beneficial Rusts; Rusts as Mycoherbicides

Although rust diseases are usually deleterious economically, they have been utilized as microbial agents for weed control. The application of rust fungi for this purpose is limited to two examples. Skeleton weed (*Chondrilla juncea*) has invaded millions of hectares of wheat and rangelands in southeastern Australia and western United States. In the former it can be a limiting factor in wheat production under certain conditions. Chondrilla rust (*Puccinia chondrillina*), indigenous to the Mediterranean, has been introduced and proved successful in control of skeleton weed in Australia. Research on this problem has recently begun in the United States, and initial results are encouraging. In the latter studies, helicopters are used to distribute urediniospore inoculum over large areas. Development of rust in skeleton weed is accompanied by defoliation and subsequent death of the plant.

Blackberry (*Rubus* spp.) was introduced by European immigrants into Chile in the nineteenth century. By 1973 it covered some 5 million hectares (12.4 million acres) of farming and grazing land in that country (Plate 1.1). *Phragmidium violaceum,* a rust of blackberry in Germany, was introduced into Chile in 1973 in several field inoculations. Within 2 years the rust covered a radius of some 70 km, and by 1976 it had spread over much of the country. Under the onslaught of infection and the reduced vigor of infected plants, blackberry is losing its competitiveness and is being replaced by more desirable species for grazing (Plate 1.2). The damge done to the plant may be summarized as the hastening of normal defoliation by 5 months; the resultant lack of lignification of stems, leading to frost damage and invasion by secondary pathogens; and up to 45% reduction of viable seed in severe infections.

REFERENCE

Ziller, W. G. 1974. The Tree Rusts of Western Canada. Canadian Forestry Service Publication 1329.

FURTHER READING

Sections 1.1-1.4

Arthur, J. C. 1929. The Plant Rusts (Uredinales). New York: John Wiley & Sons.

Carefoot, G. L., and Sprott, E. R. 1967. Famine on the Wind. Chicago: Rand McNally.

Chester, K. S. 1946. The Nature and Prevention of the Cereal Rusts as Exemplified by the Leaf Rust of Wheat. Waltham, Mass.: Chronica Botanica.

Day, P. R., ed. 1977. The genetic basis of epidemics in agriculture. Ann. N.Y. Acad. Sci. 287:1-400.

Dinus, R. J. 1974. Knowledge about natural ecosystems as a guide to disease control in managed forests. Proc. Am. Phytopathol. Soc. 1:184-190.

Green, G. J., and Campbell, A. B. 1979. Wheat cultivars resistant to *Puccinia graminis tritici* in western Canada: Their development, performance, and economic value. Can. J. Plant Pathol. 1:3-11.

Horsfall, J. G., and Cowling, E. B., eds. 1978. Plant Diseases, vol. 2. New York, San Francisco, London: Academic Press.

Chap. 1. Horsfall, J. G. and Cowling, E. B. Some epidemics man has known.

Chap. 2. Cowling, E. B. Agricultural and forest practices that favor epidemics.

Loegering, W. Q.; Johnston, C. O.; and Hendrix, J. W. 1967. Wheat rusts. In Wheat and Wheat Improvement, ed. K. S. Quisenberry and L. P. Reitz, pp. 307-335. Agronomy Series 13. American Society of Agronomy.

National Academy of Sciences. 1972. Genetic vulnerability of major crops. Washington, D.C.

Schrieber, E. 1972. Economic impact of coffee rust in Latin America. Annu. Rev. Phytopathol. 10:491–510.
Stakman, E. C., and Harrar, J. G. 1957. Principles of Plant Pathology. New York: Ronald Press.
ten Houten, J. G. 1974. Plant pathology: Changing agricultural methods and human society. Annu. Rev. Phytopathol. 12:1–11.
Wellman, F. L. 1972. Tropical American Plant Disease. Metuchen, N.J.: Scarecrow Press.

Section 1.5 (Beneficial Rusts)

Freeman, T. E., ed. 1976. Proceedings 4th International Symposium on Biological Control of Weeds. Gainesville: Univ. of Florida Press, pp. 117–121.
Oehrens, E. 1977. Biological control of the blackberry through the introduction of rust, *Phragmidium violaceum,* in Chile. FAO Plant Prot. Bull. 25:26–28.
Templeton, G. E.; TeBeest, D. O.; and Smith, R. J., Jr. 1979. Biological weed control with mycoherbicides. Annu. Rev. Phytopathol. 17:301–310.

2

Life Cycle

2.1 Spore Stages

Of all fungi, rusts have one of the most complex of life cycles. Additionally, the life cycles are varied among the genera and species; consequently, sweeping generalizations are difficult to make. Unfortunately, most introductory texts deal only with *Puccinia graminis* f. sp. *tritici,* which suggests to the beginning student that that is *the* life cycle of rust fungi.

The complete life cycle contains five successive spore stages (basidiospores, pycniospores, aeciospores, urediniospores, teliospores).[1] The last four stages are commonly designated 0, I, II, and III respectively; the basidiospores have no numeric designation (Fig. 2.1; Table 2.1).

TABLE 2.1. Stages of the complete life cycle of rust fungi

Spore	Spore-bearing structure	Numerical designation of the stage	Nuclear condition
Basidiospore	Basidium (s) Basidia (pl.)	None	1N
Pycniospore	Pycnium (s.) Pycnia (pl.)	0	1N
Aeciospore	Aecium (s.) Aecia (pl.)	I	N + N
Urediniospore	Uredinium (s.) Uredinia (pl.)	II	N + N
Teliospore	Telium (s.) Telia (pl.)	III	N + N → 2N → 1N*

* Nuclear fusion and meiosis typcally accompany germination of teliospores.

1. Numerous synonyms and variant spellings have been used for the different stages of the rust fungus life cycle. A thorough discussion of this problem is not possible in as short a book as this. Although pycniospores function as spermatia, I prefer the former term, since the structure in which they are borne is in my opinion more accurately called a pycnium than a spermagonium. The latter term should not be used to describe a structure containing both male and female components, whereas the term pycnium connotes nothing about sexuality of the components. For more detail see Hiratsuka (1973) and Savile (1976).

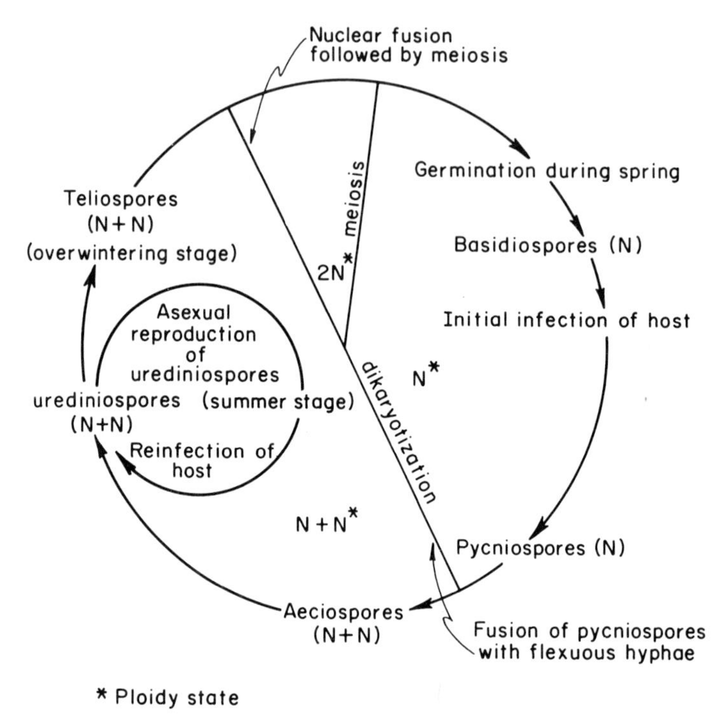

Fig. 2.1 Generalized, diagrammatic representation of the life cycle of rust fungi. All stages are not present in all rusts. In autoecious forms all stages that may be present occur on one host species. In heteroecious forms the pycnial/aecial and uredinial/telial stages occur on different host species. Teliospores are not the only overwintering stage in some rusts, and in some they are 2N, not N + N.

Many rusts lack one or more of the stages (Chap. 2.4). First we will consider the complete life cycle, beginning with the first haploid spore of the cycle, the basidiospore (Fig. 2.1; Table 2.1).

Basidiospores (Fig. 2.1) are produced upon germination of the commonly overwintering teliospores. They are borne at the apices of the four sterigmata of the basidium, the latter developing from the germinating teliospore (Figs. 2.2, 2.3). If the basidiospore lands on an appropriate host and environmental conditions are favorable, it will germinate to form a germ tube with a terminal appressorium (Fig. 2.4). A penetration peg produced from the latter typically penetrates the host cuticle and epidermis directly, giving rise to a haploid mycelial thallus within the host. In a few rusts, e.g., *Cronartium ribicola* on pine, the basidiospore germlings penetrate via stomata rather than directly into epidermal cells.

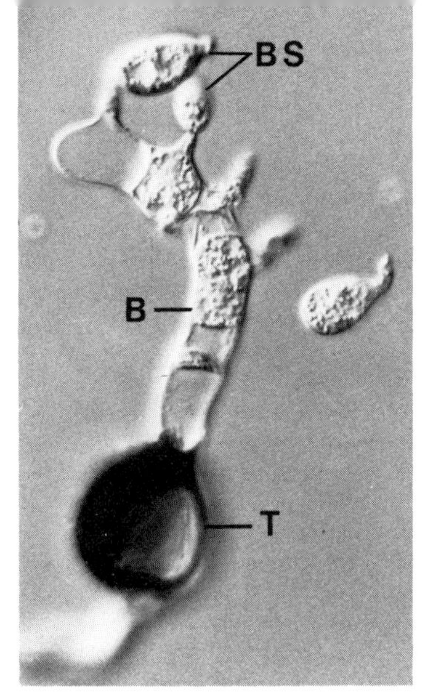

Fig. 2.2. Germination of teliospore (T) of *Uromyces phaseoli* f. sp. *typica* to produce a basidium (B). The latter produces small branches (sterigmata), at the tips of which are borne basidiospores (BS). ×960. (Courtesy R. E. Gold and K. Mendgen, University of Konstanz, West Germany)

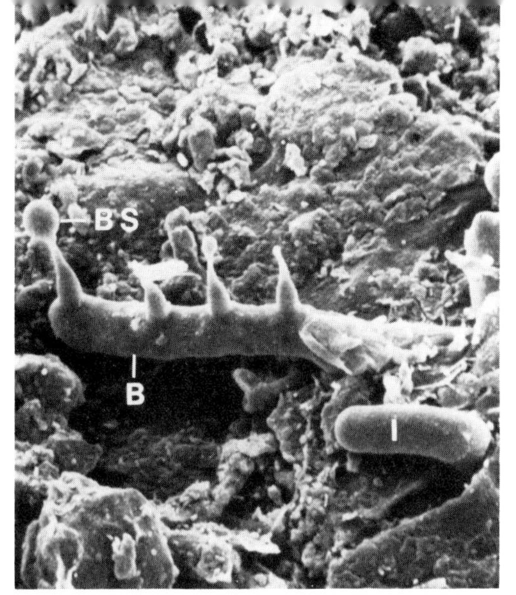

Fig. 2.3. Basidium (B) and basidiospores (BS) of *Melampsora medusae* borne on the surface of the teliospore crust; immature, emerging basidium (I). ×1000. (From Brown and Brotzman 1979, by permission of the University of Missouri Extension Division)

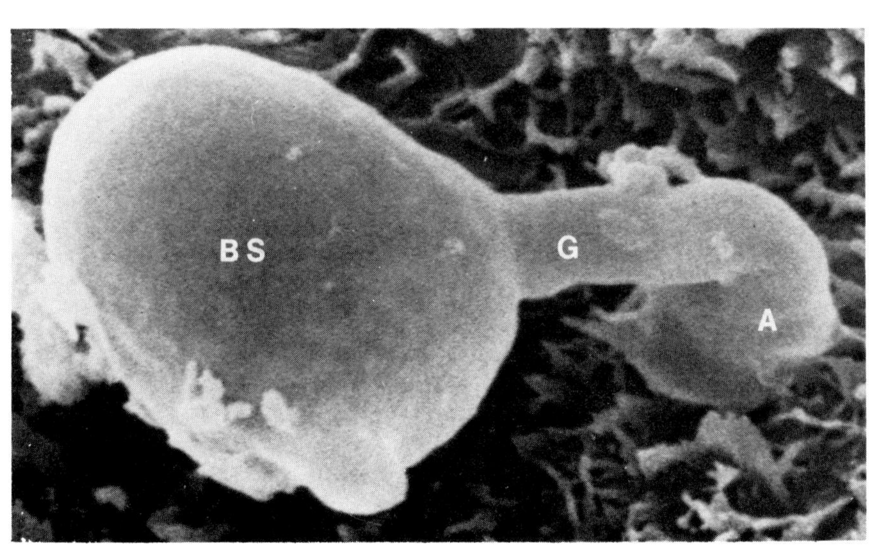

Fig. 2.4. Penetration of flax leaf by germinating basidiospore of *Melampsora lini:* basidiospore (BS); germ tube (G); appressorium at apex of germ tube (A). ×2100. (Courtesy R. E. Gold, North Dakota State University)

The haploid mycelium resulting from basidiospore infection typically remains localized in the region of host penetration, commonly in the leaves. In numerous conifer rusts, however, the mycelium grows into the woody branches and main stem of the tree. In either case, the haploid mycelium will eventually give rise to the pycnial stage, the next step in the life cycle (Table 2.1; Figs. 2.1, 2.5, 2.6; Plate 2.1).

The pycnial stage forms within 1–2 weeks after basidiospore infection in most herbaceous hosts, but it is often delayed 1–3 years in many conifer hosts. The spore-bearing structures, i.e., the pycnia, that form during this stage vary in structure and location on the host, depending on the rust. Pycnia borne on woody tissue, e.g., conifer rusts, are commonly flat masses of spore-bearing cells borne underneath the uplifted bark. On herbaceous leaves, e.g., cereal, bean, and flax rusts, the pycnia are lense- to flask-shaped structures produced beneath the epidermis (Fig. 2.6; Plate 2.1).

Pycniospores (spermatia) (Figs. 2.5b, 2.6) are borne at the tips of sporophores (the spore-bearing cells) within the pycnia. They are small, haploid, unicellular spores that function as male gametes and are in-

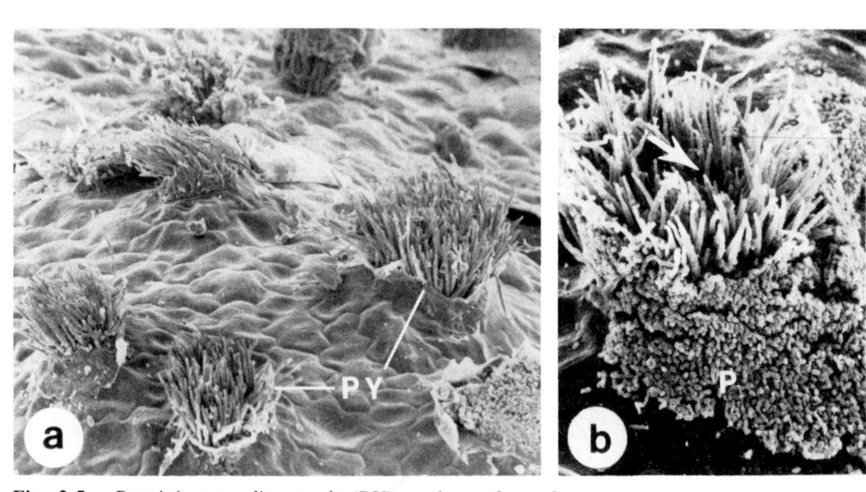

Fig. 2.5. *Puccinia recondita* pycnia (PY) on the surface of meadowrue leaf. (a) The body of each pycnium resides within the leaf tissue; only the paraphyses and flexuous hyphae rupture the leaf epidermis and extend beyond the surface. Pycniospores and honeydew were removed during specimen preparation. ×200. (Courtesy R. E. Gold, North Dakota State University). (b) An individual pycnium showing a mass of pycniospores (P) associated with the paraphyses and flexuous hyphae (arrow). ×400. (From Brown and Brotzman 1979, by permission of the University of Missouri Extension Division)

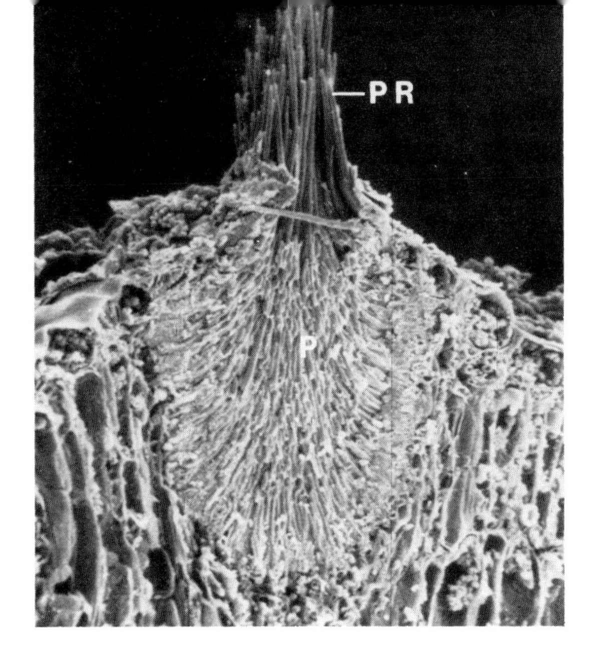

Fig. 2.6. Longitudinal fracture through a pycnium of *Puccinia recondita* showing its flask-shaped structure, pycniospores (P) produced within the pycnium, and paraphyses (PR) that extend through the host leaf epidermis. ×400. (From Brown and Brotzman 1979, by permission of the University of Missouri Extension Division)

capable of reinfecting the host. Pycniospores are borne in a viscous, sugary fluid (''honeydew'') that forms droplets at the opening of the pycnia on the host surface (Plate 2.1).

Pycnia also may contain paraphyses (periphyses or ostiolar trichomes), which aid in the rupture of the overlying host epidermis (Figs. 2.5, 2.6). They also contain ''flexuous hyphae,'' which are branched and grow up among paraphyses (Figs. 2.5, 2.6). Flexuous hyphae function as the ''female'' receptive structures with which the ''male'' pycniospores fuse to initiate the dikaryotic, i.e., N + N, phase of the life cycle (Fig. 2.1). Each pycnium and all of its male and female components are one of two mutually exclusive mating types—designated (+) and (–). Consequently, successful mating is possible only between (+) and (–) mating types, not between (–) and (–) or between (+) and (+). The ratio of (+) and (–) mating type pycnia in nature is approximately 1:1. Random matings made among pycnia are thus approximately 50% fertile and 50% sterile. Controlled matings can be made by carefully transferring honeydew and pycniospores with a fine brush from one pycnium to another. In nature this is provided by splashing raindrops, the rubbing of leaves together in the wind, or by insects attracted to the sweet-smelling honeydew.

Despite extensive research on the life cycle of rust fungi during the nineteenth century, the function of pycnia and pycniospores remained an enigma. In 1927 at the Dominion Rust Laboratory in Winnipeg, Manitoba, J. H. Craigie first proved the sexual function of pycniospores and demonstrated experimentally the transfer of pycniospores to flexuous hyphae by insects.

The aecial stage of the life cycle (Plates 2.2–2.5; Figs. 2.7–2.13) is initiated almost simultaneously with the development of pycnia in herbaceous hosts. Aecial primordia commonly form beneath the lower epidermis of the leaf and are connected to the pycnia by the haploid mycelium.

Upon fusion of pycniospores with flexuous hyphae, the nucleus of the pycniospore enters the flexuous hyphae and migrates toward the aecial primordium. In *Puccinia graminis* on barberry, this migration through the leaf requires about 24–30 hours. Once the male nuclei reach the aecial primordium, a permanent dikaryotic condition is established, with each sporogenous cell then containing one (+) and one (–) nucleus. These sporogenous cells give rise to the next spore stage of the life cycle, the aeciospores (Fig. 2.1). Since numerous pycniospores may fertilize the flexuous hyphae of a single pycnium, genetically different aeciospores may be produced within one aecium.

On nonconifer hosts the aecia commonly form on the undersides of leaves, but in some conifer rusts they form on branches or stems centrifugal to the position of the pycnia. On such woody tissues (Plate 2.4) aecium formation is often delayed a year after pycnium development, contrasted to their almost simultaneous development on leaves.

Aecial morphology is highly varied and provides significant taxonomic separation. Much of the varied morphology is associated with the nature and extent of the cell layer(s) delimiting the aecium, i.e., the peridium. Aecia are separated into five morphologic types, each having a different name.

The simplest aecium is one of the caeomoid type (caeoma, sing.; caeomata, pl.). It possesses only a rudimentary peridium. The peripheral limits of the caeoma are indistinct, being mostly fragments of ruptured host epidermis and adhering peridial cells (Plate 2.2; Fig. 2.7). Chains of aeciospores in caeomata are usually no more than 4–5 spores in length. Such aecia are typical of *Melampsora* spp. and *Phragmidium* spp.

The aecidioid aecium (aecidium, sing.; aecidia, pl.) is surrounded by a more prominent and persistent peridium. This feature provides for the "cluster cup" morphology of aecidia (Plate 2.3; Figs. 2.8, 2.9). Aeciospores commonly occur in long chains, and in most types of aecidia they are separated from each other in the chain by wedge-shaped intercalary, or "disjunctor" cells. Such aecia are typical of the genus *Puccinia*.

Peridermioid aecia (peridermium, sing.; peridermia, pl.) have a durable, extensive peridium that remains intact over the entire mass of aeciospores until maturity, at which time the peridium fragments and the aeciospores are released. Peridermia are commonly several millimeters in length and have a somewhat tonguelike to flattened, bladderlike ap-

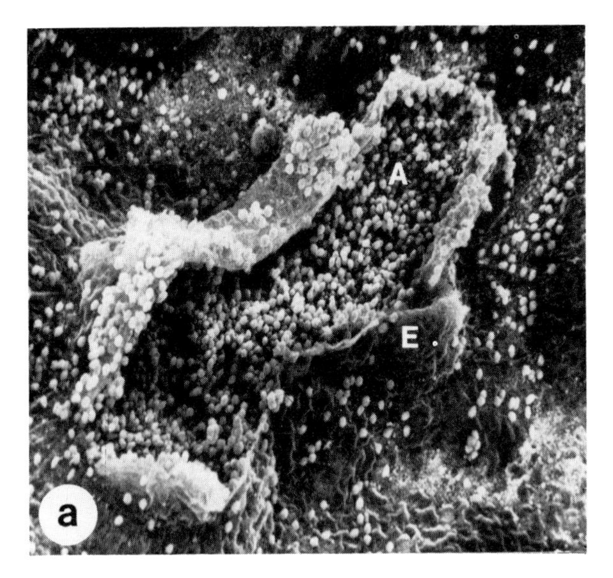

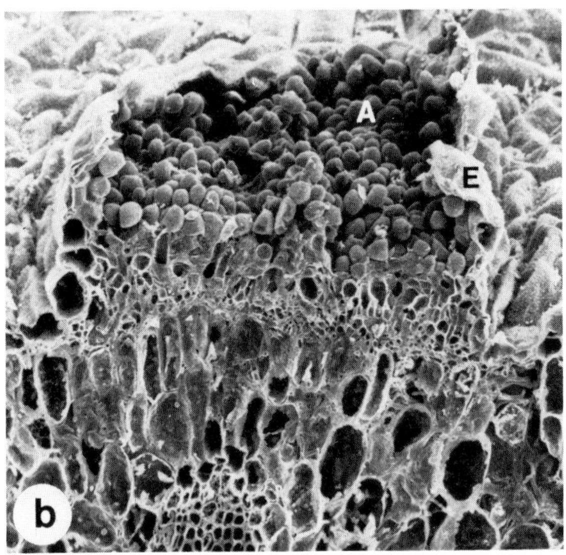

Fig. 2.7. Caeomoid aecia of *Melampsora lini.* (a) Surface view. ×75. (b) Longitudinal fracture: aeciospores (A); ruptured host epidermis (E). ×220. (From Gold and Littlefield 1979, by permission of the National Research Council of Canada)

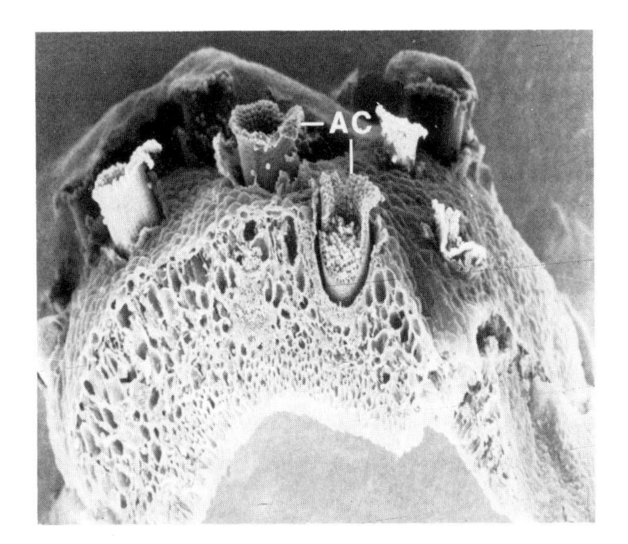

Fig. 2.8. Cross fracture of a barberry leaf showing the cuplike aecidioid aecia (AC) of *Puccinia graminis* f. sp. *tritici.* The leaf is thickened and distorted in the region of sorus development. ×40.

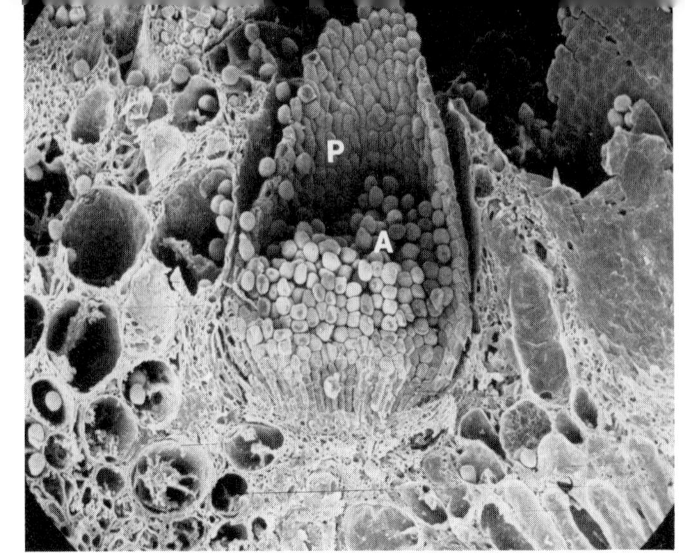

Fig. 2.9. Longitudinal fracture of an aecidioid aecium of *Puccinia recondita.* Aeciospores (A) are borne in chains, enclosed by the peridium (P) of the fungus. ×160. (From Gold et al. 1979, by permission of the National Research Council of Canada)

Fig. 2.10. Peridermioid aecia (AC) of *Melampsorella caryophyllacearum* on needles of alpine fir. Note the tubular to tonguelike shape of the former and the enlarged, chlorotic nature of the latter. (From Ziller 1974, reproduced by permission of the Minister of Supply and Services Canada)

Plate 1.1. Dense growth of blackberry bushes (arrow) in Chile. Large areas of grazing lands are overgrown by this introduced weed. (Photo by M. Lugas, courtesy F. H. Tainter, Clemson University)

Plate 1.2. Defoliation and subsequent death of blackberry in Chile by the introduced rust, *Phragmidium violaceum,* provides successful biological control. (Photo by E. Oehrens, courtesy F. H. Tainter, Clemson University)

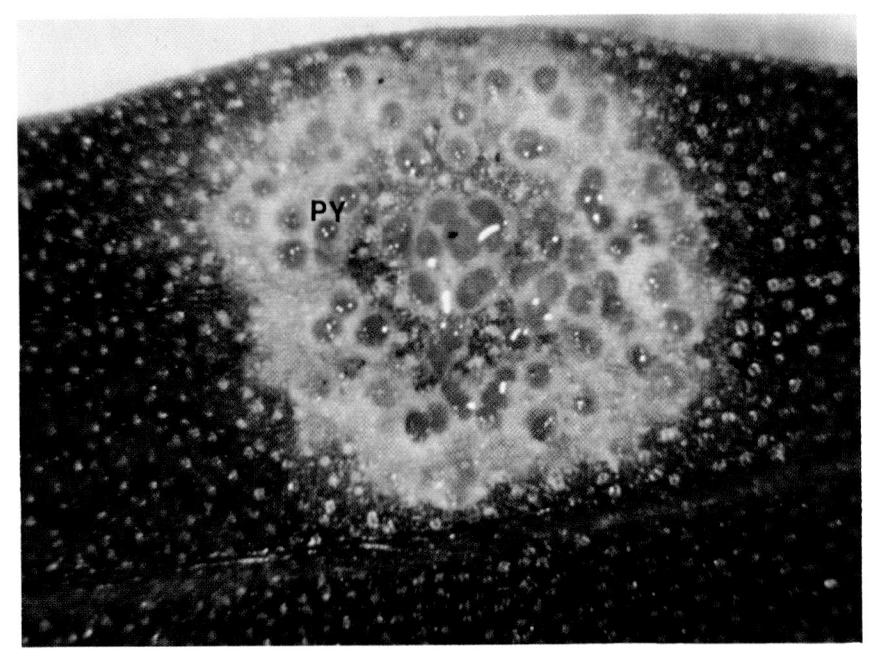

Plate 2.1. Surface of flax leaf showing a group of *Melampsora lini* pycnia (PY). Each pycnium is covered at its apex by a drop of viscous honeydew that contains pycniospores. (Courtesy R. E. Gold and G. D. Statler, North Dakota State University)

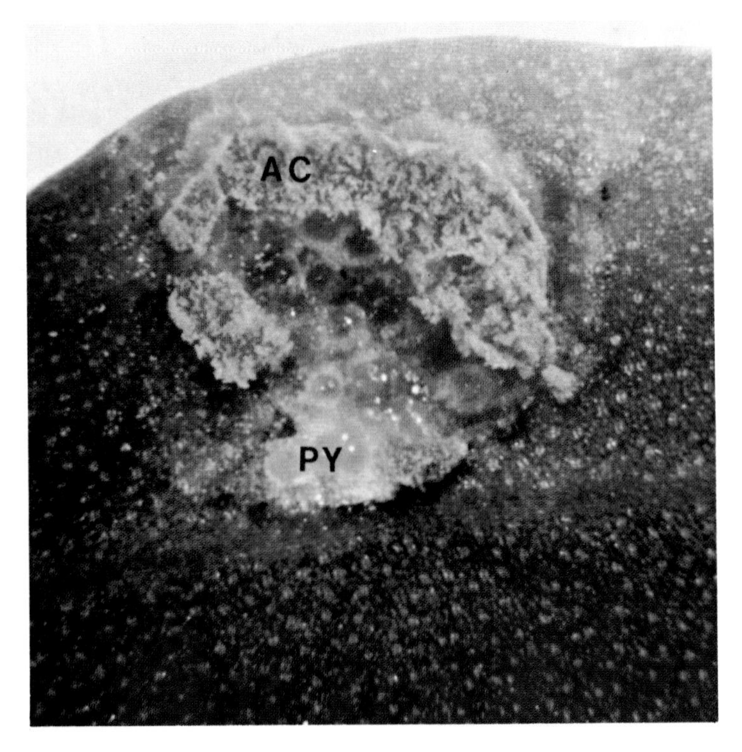

Plate 2.2. Caeomoi aecia (AC) of *Melam sora lini* adjacent to several pycnia (PY). (Courtesy R. E. Gold and G. D. Statler, North Dakota State University)

Plate 2.3. Aecidioid aecia (AC) of *Puccinia graminis* f. sp. *tritici* on underside of barberry leaf. The apices of the elongated aecia are still intact. Upon their rupture, the enclosed aeciospores will be exposed to the air, and the aecia will assume their typical cup-shaped morphology as in Fig. 2.8. ×5. (Courtesy A. P. Roelfs, U.S. Department of Agriculture)

Plate 2.4. (Lower left) Peridermioid aecia (AC) of *Cronartium ribicola* on trunk and branches of eastern white pine. (Courtesy T. Nicholls, U.S. Department of Agriculture)

Plate 2.5. (Above) Roestelioid aecia (AC) of *Gymnosporangium bethelii* on leaves of hawthorn. (From Ziller 1974, reproduced by permission of the Minister of Supply and Services Canada)

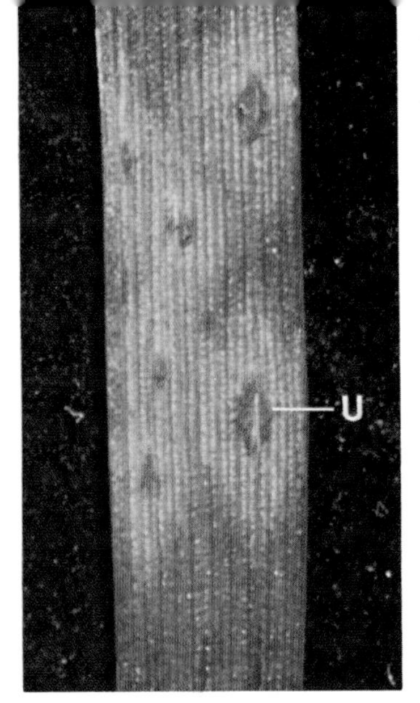

Plate 2.6. Uredinia (U) of *Puccinia graminis* f. sp. *tritici* on wheat leaf. Note the region of chlorosis surrounding each uredinium. ×3. (Courtesy A. P. Roelfs, U.S. Department of Agriculture)

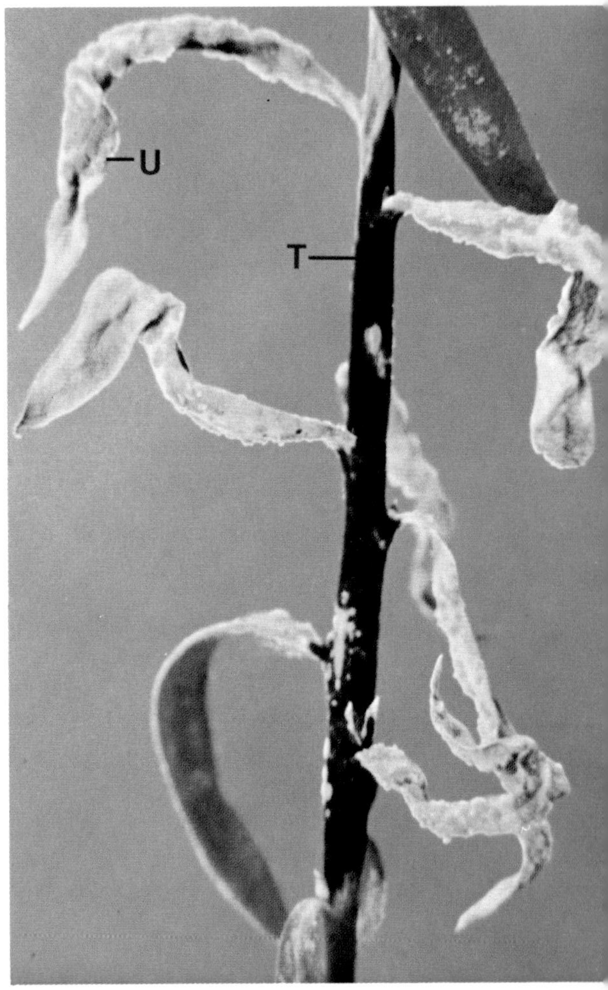

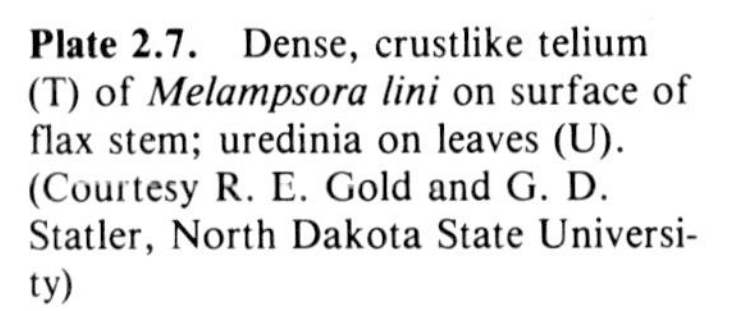
Plate 2.7. Dense, crustlike telium (T) of *Melampsora lini* on surface of flax stem; uredinia on leaves (U). (Courtesy R. E. Gold and G. D. Statler, North Dakota State University)

Plate 3.1. Telial gall of *Gymnosporangium juniperi-virginianae* on juniper. The central mass (here, concealed) consists of host and fungal cells; the surface extensions consist primarily of aggregated teliospore pedicels. (Courtesy T. H. Nicholls, U.S. Department of Agriculture)

Fig. 2.11. Peridermioid aecium (AC) of *Coleosporium senecionis* on pine needle. ×40. (From Littlefield and Heath 1979, by permission of Academic Press)

pearance (Plate 2.4; Figs. 2.10, 2.11). Peridermia are restricted to conifer hosts, occurring on either woody tissue (e.g., *Cronartium* spp.) or needles (e.g., *Coleosporium* spp. and *Melampsora* spp.).

Roestelioid aecia (roestelium, sing.; roestelia, pl.) are commonly the most extensively developed form of aecia and may extend up to 1 cm in length (Plate 2.5; Figs. 2.12, 2.13). They are restricted to the genus *Gymnosporangium*. Here again the peridium provides the outer, structural boundry of the sorus and surrounds the long chains of aeciospores borne within the columnar structure of the roestelium. The peridium may open in a variety of ways to release the aeciospores. In *G. globosum* the peridium disintegrates in a basipetal direction, which allows release of aeciospores progressively lower in the chains. In *G. juniperi-virginianae* (Fig. 2.13) the peridium consists of long files of hinged cells that curl outward in low humidity to expose the spore chains and curl inward during high humidity, which limits spore release.

Fig. 2.12. Roestelioid aecia of *Gymnosporangium bethelii* on fruit and leaf of hawthorn. (From Ziller 1974, reproduced by permission of the Minister of Supply and Services Canada)

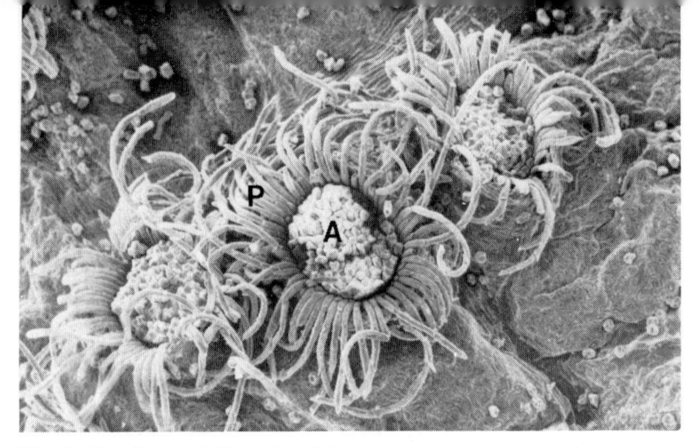

Fig. 2.13. Roestelioid aecia of *Gymnosporangium juniperi-virginianae* on apple leaf: aeciospores (A); files of peridial cells (P). ×80. (From Littlefield and Heath 1979, by permission of Academic Press)

The remaining type of aecium, the uredinioid aecium (uraecium, sing.; uraecia, pl.), is somewhat a "morphologic hybrid" between the "true" aecium and the next stage of the life cycle, the uredinium. The peridium is absent or greatly diminished, and the pustule is quite diffuse and indistinct. The spores are borne on stalks, not in chains. Morphologically the structure is that of uredinia and urediniospores,[2] but functionally the spores of a uraecium originate and behave like aeciospores, as described below. Uraecia are characteristic of the genera *Pileolaria, Prospodium,* and *Uropyxis,* all of which are generally restricted to warmer regions.

The surface of aeciospores is usually ornamented with small knoblike structures. These may be smooth or annulate in outline, depending on the rust (Fig. 2.14).

Aeciospores reinfect the same host species, if autoecious, or the alternate host species, if heteroecious (Chap. 2.2), but in either case, that infection leads to the production of the next step of the life cycle, the uredinial stage. Thus aeciospores are not "repeating" spores that produce successive generations of aeciospores. Aeciospore infection occurs typically through host stomata. Mycelium from the dikaryotic spores forms a dikaryotic thallus in the host tissue.

The uredinial stage usually forms within 1–2 weeks after infection by aeciospores. Uredinia are circular to oblong sori that bear stalked or catenulate urediniospores and may be surrounded by a peridium, by paraphyses, or by only the ruptured host epidermis (Plate 2.6; Fig. 2.15). The dikaryotic urediniospores typically have spinelike ornaments over their entire surface (Fig. 2.16). They can reinfect the host to produce

2. Uredinia and urediniospores are often termed uredia and urediospores or uredospores. The three latter terms, however, are etymologically incorrect, as explained by Savile (1968), and the correct terms, uredinium (sing.), uredinia (pl.), and the derivatives, uredinial, urediniospore, etc., are accepted by this author.

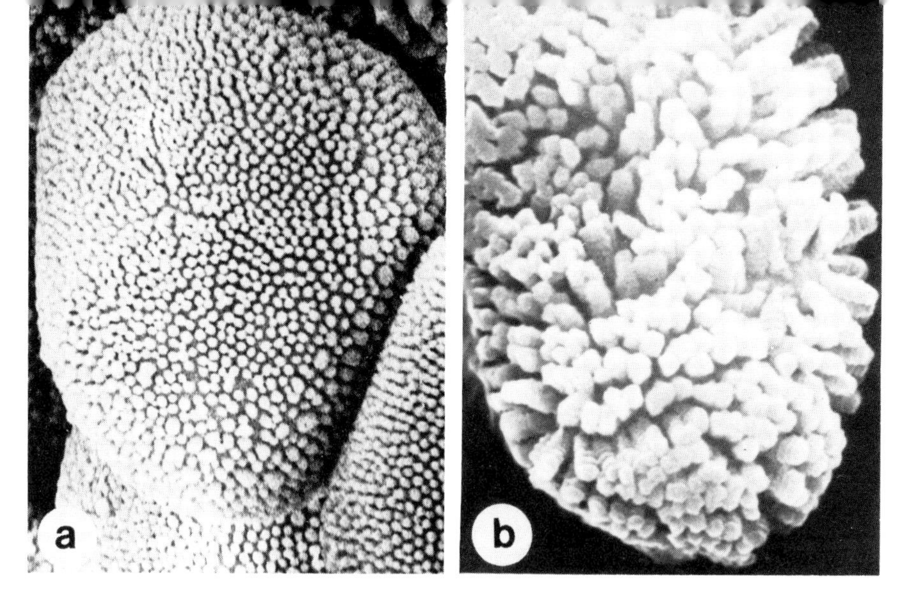

Fig. 2.14. Common types of aeciospore surface ornamentation. (a) Smooth, knoblike ornaments of *Puccinia* spp. ×3100. (b) Annulated knobs of *Cronartium* spp. ×4000. (From, respectively, Gold et al. 1979 and Hiratsuka 1971, by permission of the National Research Council of Canada)

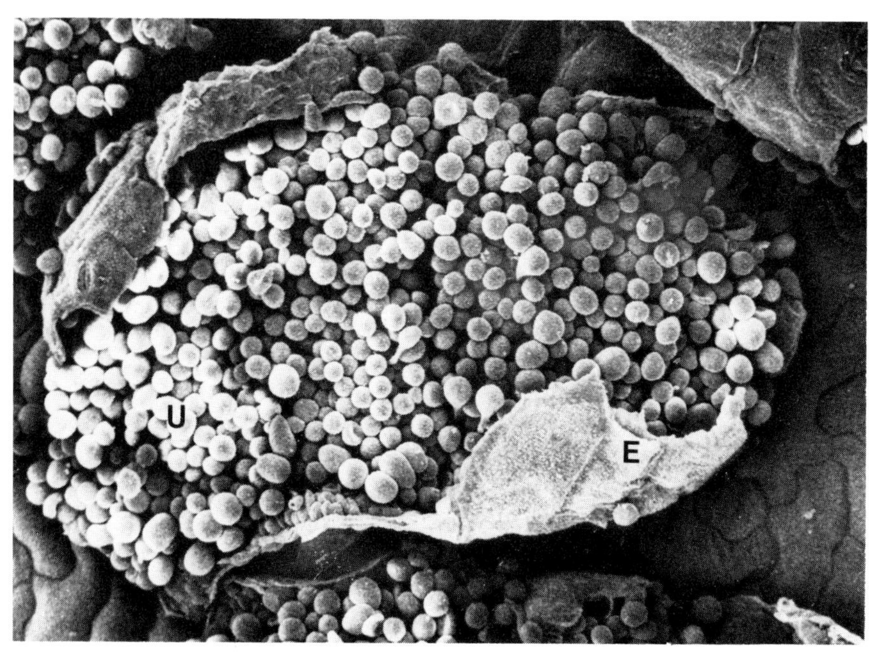

Fig. 2.15. Uredinium of *Melampsora lini* on flax leaf: urediniospores (U); ruptured host epidermis (E). ×250. (Courtesy Z. M. Hassan, North Dakota State University)

several generations of uredinia; consequently, they are referred to as "repeating" spores (Fig. 2.1). The rapid, widespread buildup of the repeating uredinial stage is responsible for epidemics of cereal, bean, coffee, and other rusts over large areas of crop monoculture.

A modified form of urediniospore, the amphispore, occurs in some rust species. Amphispores are capable of surviving extended periods of unfavorable conditions that would be lethal to most urediniospores. They have thicker and more deeply pigmented walls than ordinary urediniospores and can be easily mistaken for teliospores.

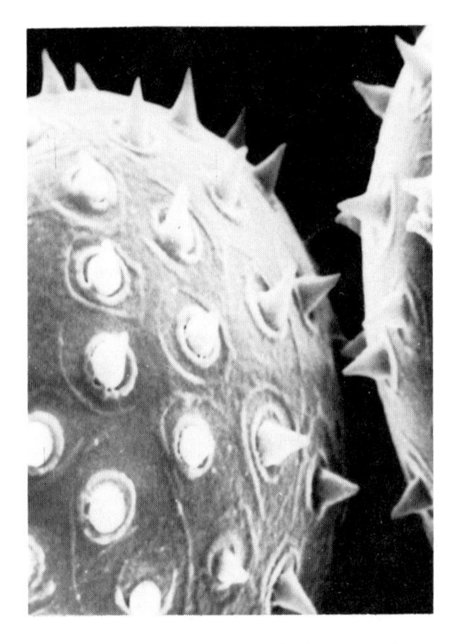

Fig. 2.16. Urediniospores of *Uromyces phaseoli* f. sp. *typica* showing the spinelike (echinulate) ornamentation typical of most urediniospores. ×6100. (Courtesy J. R. Venette, from Littlefield and Heath 1979, by permission of Academic Press)

Eventually, the infected plants begin to produce the telial stage of the life cycle (Figs. 2.1, 2.17–2.20; Plate 2.7). Initially, teliospores of some rusts may begin to form within uredinia; thus teliospores and urediniospores are together in one sorus. Later, only teliospores are formed, and no urediniospores are associated with them. Telium morphology is quite varied and provides the major criterion for classification of the rusts (see Chap. 3). Except in cases where dormant mycelium in host tissue is the overwintering structure, rusts commonly overwinter as dormant, thick-walled teliospores. Teliospores exhibit a vast array of ornamentation types, being smooth, spined, verrucose, reticulate, or punctate, to mention but a few of the variants (see Figs. 3.2, 3.3). Also, depending on the genus, the teliospores may be stalked or sessile, individual or united, or single celled or multicellular (see Fig. 3.1).

Fig. 2.17. Telia of *Puccinia graminis* f. sp. *tritici* on stems (peduncles) of wheat. (Courtesy A. P. Roelfs, U.S. Department of Agriculture)

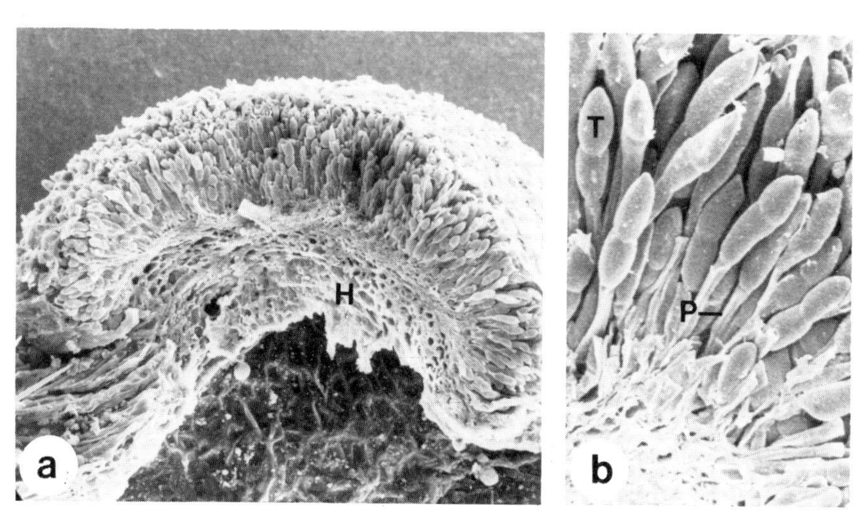

Fig. 2.18. *Puccinia malvacearum* on hollyhock. (a) Cross fracture through telium; host tissue (H). ×60. (b) Enlargement showing the two-celled teliospores (T) borne on pedicel cells (P). ×180.

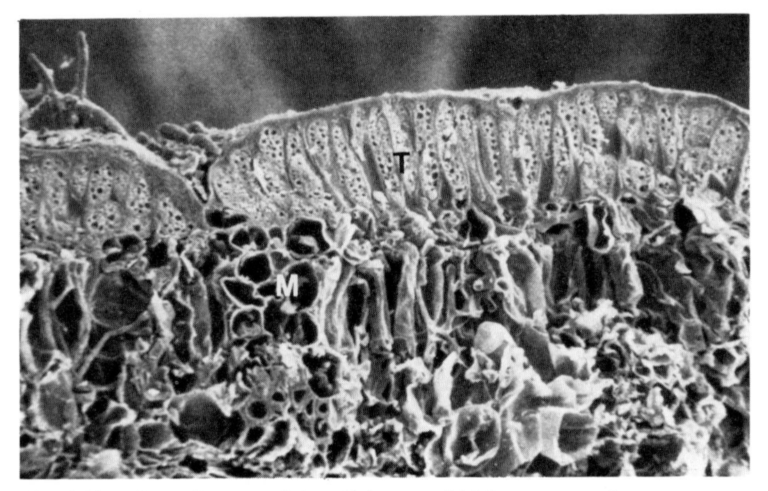

Fig. 2.19. Cross fracture of the telial crust of *Melampsora medusae* on cottonwood leaf. The palisade of teliospores (T) forms between the host mesophyll (M) and the epidermis, which has already sloughed off in this preparation. ×300. (From Brown and Brotzman 1979, by permission of the University of Missouri Extension Division)

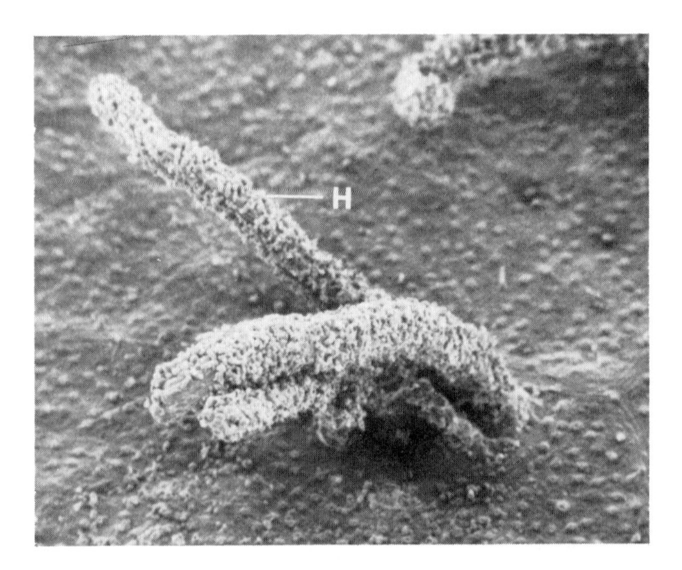

Fig. 2.20. Telial horns (H) of *Cronartium fusiforme* on oak leaf. The horns consist of aggregated teliospores that bear basidia and basidiospores over their surface as the teliospores germinate. ×75. (From Brown and Brotzman 1979, by permission of the University of Missouri Extension Division)

Upon germination, teliospores produce typical basidia and basidiospores (Figs. 2.1–2.3, 2.21). Accompanying germination, the (+) nucleus and the (−) nucleus in each teliospore fuse to form an ephemeral diploid stage. This is soon followed by meiosis, resulting in four haploid nuclei, two of the (+) mating type and two of the (−) mating type. The four nuclei then migrate into each of the four respective basidiospores produced on the basidium (Figs. 2.2, 2.3). The life cycle is thus completed.

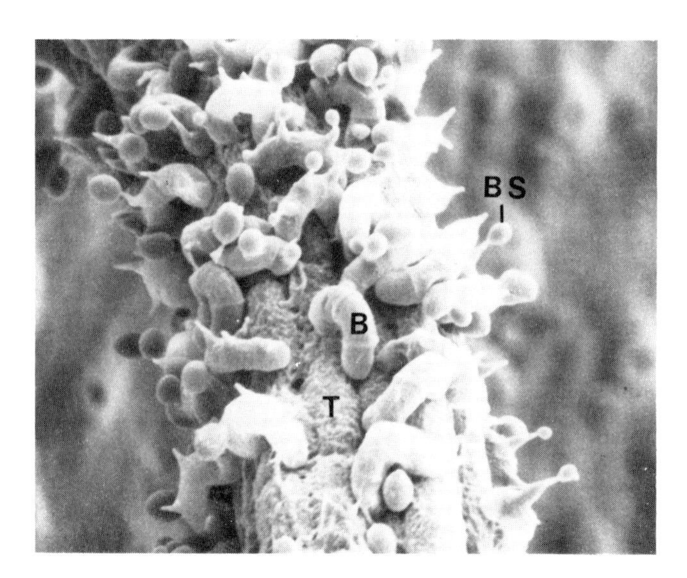

Fig. 2.21. Enlargement of a portion of Figure 2.20 showing basidia (B) arising from germinated teliospores (T) in the telial horn; basidiospore (BS). ×600. (From Brown and Brotzman 1979, by permission of the University of Missouri Extension Division)

2.2 Autoecism and Heteroecism

Rusts capable of completing their entire life cyle on one host species are termed autoecious (Gr. *autos* = self, the same + *oikos* = home, host). Examples of such rusts include *Melampsora lini* (on flax), *Phragmidium* spp. (on rose), *Puccinia asparagi* (on asparagus), and *Uromyces phaseoli* (on bean).

Rusts that require two host species for the completion of their life cycle are termed heteroecious (Gr. *heteros* = other, different + *oikos* = home). The separation of the life cycle on the respective hosts occurs immediately after the aecial stage (Fig. 2.1). Thus *Puccinia graminis* occurs as pycnia and aecia on barberry and as uredinia and telia on various grasses. Similarly, *Melampsora medusae* forms pycnia and aecia on larch, but its uredinia and telia are on poplar or cottonwood. The host species of heteroecious rusts are always very different taxonomically, e.g., the fungus will alternate between ferns and conifers, gymnosperms and angiosperms, or monocots and dicots. With few exceptions, e.g., *Gymnosporangium* spp. on apple and juniper, the pycnial/aecial host is the more primitive evolutionarily compared to the uredinial/telial host. The term "alternate host" is used to denote either of the two unlike hosts of a heteroecious rust, but it often refers to the one of lesser economic importance. It may represent either the pycnial/aecial or the uredinial/telial stage host.

Sometimes the alternate hosts of presumably heteroecious rusts are not known. For example, in coffee rust (*Hemileia vastatrix*), only the uredinial and telial stages are known. Basidiospores formed upon germination of teliospores will not reinfect coffee. Presumably another host exists for the pycnial and aecial stages, but it has yet to be discovered.

The fact that one rust species may possess four different spore stages, often on two different hosts, provided monumental challenges to early investigators of the rusts. This clearly defined variable morphology, referred to as pleomorphism, was first shown by the Tulasne brothers (one a physician-botanist, the other a lawyer) in the 1850s. Anton DeBary, the father of modern plant pathology, proved the heteroecious nature of the wheat stem rust fungus in 1865, providing at last a scientific explanation to the centuries-old knowledge of farmers that wheat rust was more severe in the vicinity of barberry bushes.

2.3 Imperfect (asexual) Forms

Sometimes only the pycnial/aecial or just the uredinial stages are known to exist in nature, and an "imperfect" (asexual) name must be given the fungus until the "perfect" (sexual, or telial, stage) can be found and identified as the same species as the "imperfect" species in question. For example, many present-day *Puccinia* spp. and *Melampsora* spp. (identified by their telial stage morphoplogy) were tentatively designated as *Uredo* spp., based on the presence of only the uredinium at the time of their first description. Similarly, the "imperfect" genera, *Caeoma, Aecidium, Roestelia,* and *Peridermium,* were defined solely on the basis of morphology of their respective aecial stages (see pp. 20–23). When the connection was made between the pycnial/aecial and the respective telial stages, the generic name was changed. For example, *Roestelia hyalina* ("imperfect" name) became *Gymnosporangium hyalinum* ("perfect" name), and *Peridermium stalactiformi* ("imperfect" name) became *Cronartium coleosporioides* ("perfect" name) based on the connection between the aecial and the telial stages. Similarly, *Uredo lini* ("imperfect" name) became *Melampsora lini* ("perfect" name), based on the connection between the uredinial and the telial stages.

2.4 Abbreviated Forms

Not all rusts possess all five spore stages in their life cycle. The teliospore is the only stage that is essential for sexual reproduction, since meiosis occurs either there or in the newly formed basidium as teliospore germination ensues. Complex nomenclatural systems have been devised to designate the variant life cycles, depending on which spore stage(s) is missing. Basically, however, there are three fundamental types: (1) macrocyclic—when all spore stages are present; (2) demicyclic—when the uredinial stage is absent; and (3) microcyclic—when only the pycnial and telial or just the telial stages are present (basidiospores can reinfect the host in all microcyclic rusts). The microcyclic rusts (e.g., *Puccinia malvacearum* on hollyhock) are all necessarily autoecious. The macrocyclic and demicyclic rusts may be either autoecious or heteroecious.

REFERENCES

Brown, M. F., and Brotzman, H. G. 1979. Phytopathogenic Fungi: A Scanning Electron Microscopic Survey. Columbia: Univ. of Missouri Extension Division.

Craigie, J. H. 1927. Discovery of the function of pycnia of the rust fungi. Nature (London) 120:765-767.

Gold, R. E., and Littlefield, L. J. 1979. Ultrastructure of the telial, pycnial and aecial stages of *Melampsora lini*. Can. J. Bot. 57:629-638.

Gold, R. E.; Littlefield, L. J.; and Statler, G. D. 1979. Ultrastructure of the pycnial and aecial stages of *Puccinia recondita*. Can. J. Bot. 57:74-86.

Hiratsuka, Y. 1971. Spore surface morphology of pine stem rusts of Canada as observed under a scanning electron microscope. Can. J. Bot. 49:371-372.

______. 1973. The nuclear cycle and the terminology of spore stages in Uredinales. Mycologia 65:432-443.

Littlefield, L. J., and Heath, M. C. 1979. Ultrastructure of Rust Fungi. New York, London: Academic Press.

Savile, D. B. O. 1976. Evolution of the rust fungi (Uredinales) as reflected by their ecological problems. In Evolutionary Biology, vol. 9, ed. M. K. Hecht, W. C. Steere, and B. Wallace, pp. 137-207. New York: Plenum.

Ziller, W. G. 1974. The Tree Rusts of Western Canada. Canadian Forestry Service Publication 1329.

FURTHER READING

Alexopoulos, C. J., and Mims, C. W. 1979. Introductory Mycology, 3rd ed. New York, London, Sydney: John Wiley & Sons.

Cummins, G. B. 1959. Illustrated Genera of Rust Fungi. Minneapolis: Burgess.

3

Taxonomy

3.1 Introduction: Suprafamily Characteristics

As with any manufactured concept, taxonomy frequently changes. As more is learned about the genetic, morphological, cytochemical, and other properties of organisms, their hypothesized interrelationships often shift. Thus taxonomy is a dynamic science, notwithstanding widely held opinions to the contrary. The shifting of relationships and changing of names are often frustrating to the student who has committed to memory a given set of names and relationships, but such is the nature of science. Also, considering human nature, it is unlikely we will ever achieve a classification scheme that will remain unaltered.

Traditionally, based on nineteenth-century taxonomic systems, the fungi were divided into several classes, with the rusts belonging to class Basidiomycetes. More recent systems have constructed the division (phylum) Eumycota for the true fungi, which consists of several subdivisions. The subdivision Basidiomycotina is an example and is equivalent to the traditional class Basidiomycetes of the older systems. Separations within the subdivision Basidiomycotina vary with different workers. Following is the classification scheme for the basidiomycetes as presented by Alexopoulos and Mims (1979).

Subdivision Basidiomycotina
- Class Basidiomycetes
  - Subclass Teliomycetidae
    - Order Uredinales (rusts)
    - Order Ustilaginales (smuts and basidiomycetous yeasts)
  - Subclass Phragmobasidiomycetidae (3 orders, including jelly fungi and relatives)
  - Subclass Holobasidiomycetidae (5 orders, including agarics, boletes, polypores, puffballs, and stinkhorns)

Morphologically, the subdivision Basidiomycotina is characterized by the presence of 4 one-celled basidiospores borne on a basidium.

Usually these spores are produced at the apices of pointed sterigmata, which arise from the surface of the basidium (see Figs. 2.2, 2.3, 2.21). The order Uredinales is characterized, among other features, by having such septate basidia with sterigmata, being formed upon germination of the often thick-walled and often dormant teliospore.

3.2 Family and Genera Characteristics

The order Uredinales is divided into three families, based primarily on the manner in which the teliospores germinate and how they aggregate to form the telia sorus. The following key to the families is modified from Alexopoulos and Mims (1979).

A. Teliospore forming a septate mycelium upon germination
- B. Teliospores free or variously united but never in the form of layers or crusts . . . . Pucciniaceae (Fig. 3.1a; see also Fig. 2.18)
- BB. Teliospores laterally united into layers, crusts, or columns . . . . Melampsoraceae (Fig. 3.1b; see also Plate 2.7; Fig. 2.19)

AA. Teliospore, during germination, converts into a basidium by formation of septa; no external basidium formed . . . . Coleosporiaceae (Fig. 3.1c)

The families are separated into genera primarily on the morphology of telia and teliospores. However, uredinia and urediniospores or aecia and aeciospores are used for identification of "imperfect" rusts (see Chap. 2.4) or sometimes when specimens lack the telial stage. Also, because of the parasitic specialization of rusts on their hosts, correct identification of the host is an invaluable aid in identifying a rust fungus.

Although some 130 genera of rusts are often recognized, the farmer, forester, or nursery grower will probably encounter less than a dozen or so genera outside the tropics. Following is a brief description of some such genera, with the emphasis placed on telial morphology—the stage used primarily in generic separation.

3.2.1 FAMILY PUCCINIACEAE

***Puccinia* spp. (Fig. 3.1a; see also Fig. 2.18).** This genus, the largest of the rusts, with some 3000–4000 species, occurs on many families of angiosperms worldwide. Heteroecious forms often have grasses as their

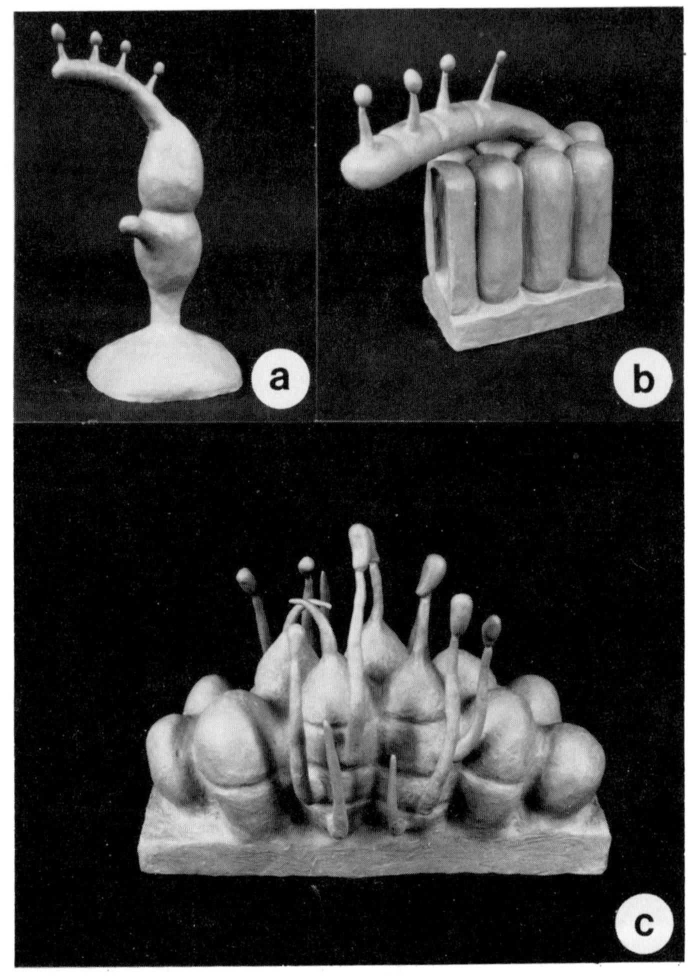

Fig. 3.1. Models showing comparative morphology of telia and associated germination structures in the three families of Uredinales. See key and family descriptions, p. 33. (a) Pucciniaceae. (b) Melampsoraceae. (c) Coleosporiaceae.

uredinial/telial stage host. Teliospores are generally two-celled, are borne singly on pedicels, and may be smooth surfaced or variously ornamented. As mentioned previously, many important plant diseases are caused by species of this genus, e.g., black stem rust of cereals and grasses (*P. graminis*), crown rust of oats and grasses (*P. coronata*), stripe or yellow rust of wheat and grasses (*P. striiformis*), common maize rust

(*P. sorghi*), tropical maize rust (*P. polysora*), sugarcane rust (*P. melanocephala*), sunflower rust (*P. helianthi*), asparagus rust (*P. asparagi*), snapdragon rust (*P. antirrhini*), groundnut or peanut rust (*P. arachidis*), and many others.

***Uromyces* spp. (Fig. 3.2).** This second largest genus of rusts contains some 600 species on a wide variety of angiosperm hosts worldwide. Common hosts include the Gramineae, Compositae, and Leguminosae. The genus is separated from *Puccinia* only in having unicellular teliospores and is maintained separately only for convenience—by "gentlemen's agreement." Economically important examples are bean rust (*U. phaseoli* f. sp. *typica*), alfalfa or lucerne rust (*U. striatus*), clover rust (*U. trifolii* f. sp. *repentis*), carnation rust (*U. dianthi*), beet rust (*U. betae*), and others.

***Phragmidium* spp. (Fig. 3.3).** About 60 species have been described, nearly all in temperate zones. All are autoecious and occur on species of the Rosaceae. Teliospores are borne singly on pedicels and commonly are several-celled. Various species occur on wild and cultivated rose, blackberry, *Potentilla* spp., and other members of the rose family.

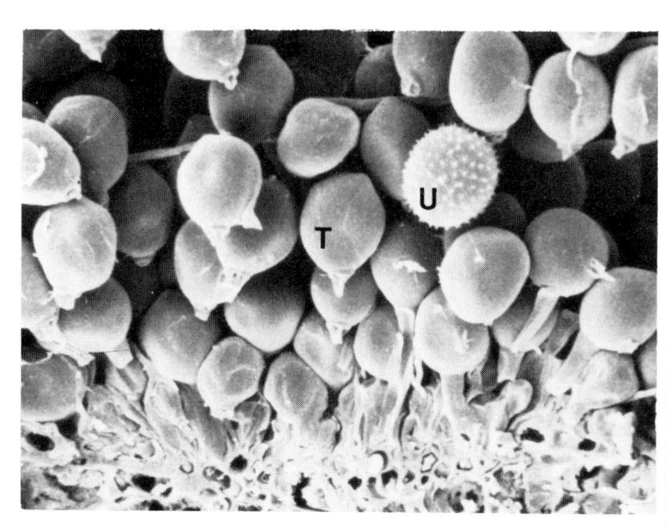

Fig. 3.2. Single-celled, smooth-surfaced teliospores (T) of *Uromyces trifolii* on red clover; urediniospore (U). ×650. (From Brown and Brotzman 1979, by permission of the University of Missouri Extension Division)

Fig. 3.3. Teliospore of *Phragmidium imitans.* These multicelled spores are typically covered with closely spaced, irregularly shaped ornaments. ×600. (From Littlefield and Heath 1979, by permission of Academic Press)

Gymnosporangium **spp. (Plate 3.1; Fig. 3.4).** This typically heteroecious, rather small genus (some 50 species) is easily recognized by its large, bright orange telial galls, which are visible on infected juniper hosts. Such galls appear following spring and summer rains. Morphologically, the teliospores are similar to *Puccinia* in being two-celled and borne singly on pedicels (Fig. 3.4). However, the pedicels are exceedingly long, highly hygroscopic, and gelatinous; they form long "telial horns" over the surface of telial galls (Plate 3.1). Each horn is composed primarily of gelatinized teliospore pedicels. The genus is typically temperate and generally heteroecious. It is unique in that the telial stage is on the gymnospermous host and the pycnial/aecial stage is on the angiospermous host. It produces distinctive roestelioid aecia on the latter (see Plate 2.5; Figs. 2.12, 2.13). *G. juniperi-virginianae* is an important pathogen of apple (aecial stage) as well as ornamental junipers (telial stage).

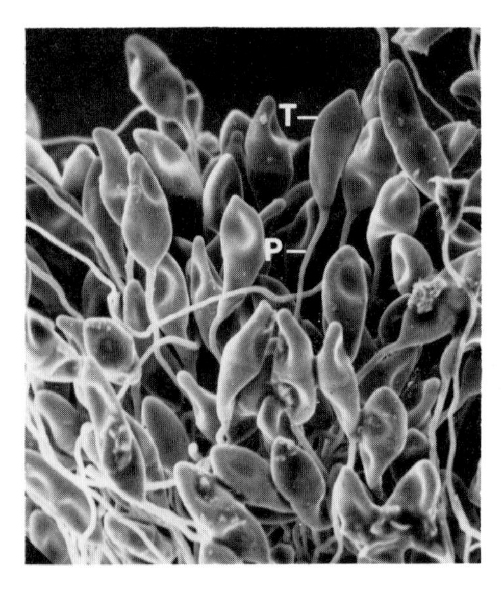

Fig. 3.4. Teliospores (T) of *Gymnosporangium junipera-virginianae* on the surface of telial horns (see Plate 3.1). Note the long teliospore pedicels (P). ×360.

3.2.2 FAMILY MELAMPSORACEAE

Melampsora **spp. (Fig. 3.1b; see also Plate 2.7; Fig. 2.19).** Some 80 northern temperate, autoecious or heteroecious species of this genus have been described. Its characteristic feature is the subepidermal crust of sessile, laterally adherent, single-celled teliospores that forms near the surface of the infected host. Important examples are flax rust (*M. lini,* autoecious) and the heteroecious species that produce uredinia and telia on various poplars and willows and pycnia and aecia on larch and fir species (e.g., *M. medusae*).

***Cronartium* spp. (see Plate 2.4; Figs. 2.20, 2.21).** Consisting of some 20 species, this genus is entirely heteroecious, with uredinial/telial stages on various dicots and pycnial/aecial stages on stems and cones of pines. Telial columns (see Figs. 2.20, 2.21) consist of nonpedicellate teliospores aggregated laterally and longitudinally into fascicles. These contrast with telial horns of *Gymnosporangium* spp., which consist of aggregated teliospore pedicels. White pine blister rust (*C. ribicola*) and southern pine fusiform gall rust (*C. fusiforme*) are two extremely important diseases caused by members of this genus (see Chap. 1.2). Alternate hosts in the uredinial/telial stages of these two rusts, respectively, include currant and oak.

3.2.3 FAMILY COLEOSPORIACEAE

***Coleosporium* spp. (Fig. 3.1c).** Although some 80 species have been described, many of them are doubtfully distinct morphologically. Most species are heteroecious, with the pycnia and aecia on needles of various pines and the uredinia and telia on various mono- and dicot hosts. Compositae is the most common host family for the uredinial/telial stage. As characteristic for the family, the teliospores of this genus do not form external basidia upon germination as in the two other rust families. Rather, the teliospore is converted into a four-celled basidium by the formation of three septa. Each resulting cell gives rise to an elongated sterigma on which the basidiospore is borne (Fig. 3.1c). Although yellowing and premature dropping of pine needles often result from *Coleosporium* infection, injury to the tree or economic losses are generally not severe.

3.3 Species and *Formae Speciales*

The some 4000–5000 species of rusts are separated primarily on the bases of spore morphology and specialization on their hosts. For example, in the uredinial/telial stage, *Puccinia graminis* will infect only certain grasses, *P. arachidis* only certain legumes, *P. helianthi* only sunflowers, etc.

Rust species (identified morphologically) that attack more than one host species are sometimes divided into more specialized categories, designated "varieties" and "specialized forms," or the Latin, *forma specialis* (*formae speciales,* pl.). This degree of specialization within species was first described by Ericksson and Henning in Sweden in the 1890s and has since been applied by other workers to numerous phytopathogenic fungi. For example, *Puccinia graminis* can be divided into:

P. graminis f. sp. *avenae* on oats and certain other grasses
P. graminis f. sp. *phleipratensis* on timothy and certain other grasses
P. graminis f. sp. *poae* on bluegrass
P. graminis f. sp. *secalis* on rye and certain other grasses
P. graminis f. sp. *tritici* on wheat and barley

Other *formae speciales* of this species have also been described. Similarly, *Uromyces phaseoli* f. sp. *typica* attacks beans, while *U. phaseoli* f. sp. *vignae* attacks cowpeas.

Slight morphologic differences, e.g., dimensions of urediniospores, have been described for various *formae speciales,* but variations in spore size within the respective categories make this a questionable criterion for identification. Identification of *formae speciales* is best made by determination of host specificity.

3.4 Physiologic Races

Host specialization to the level of cultivar (variety) commonly occurs in rusts. It may occur within species (e.g., *Melampsora lini* on flax) or within *formae speciales* (e.g., *Puccinia graminis* f. sp. *tritici* on wheat). It is manifested by some strains of the rust attacking only certain, and not other, cultivars of the host. Such variants of the pathogen are called physiologic (pathogenic) races and traditionally have been assigned a number in the order of their recognition. The discovery of this phenomenon by E. C. Stakman at the University of Minnesota in 1916 is a benchmark in the history of plant pathology. It continues to be a principle on which breeding for resistance to many plant diseases is based.

A thorough discussion of the techniques for identification of physiologic races is beyond the scope of this book. Part of the problem lies with the fact that a "standard" system was developed by Stakman and was used for over 50 years. During that time more than 300 physiologic races of *Puccinia graminis* f. sp. *tritici* were described and volumes of associated literature were published. However, a totally new system, with slight variations, has been developed by Canadian, U.S., and Australian workers in the past decade. It has almost entirely replaced the original Stakman system, but as yet it is essentially unknown to workers not involved with rust race identification and associated plant breeding.

Stakman established, after extensive testing, a set of 12 "differential varieties" (cultivars) that differ in their susceptibility to different physiologic races of *P. graminis* f. sp. *tritici.* Likewise, each physiologic race has a specific spectrum of the differential varieties it will attack; the responses are recorded and grouped into one of three "reaction classes": resistant, susceptible, or heterogeneous (mesothetic). Within the first two reaction classes are six numbered "infection types," which are designated 0, 0; (zero fleck), 1 and 2 in the resistant class, and 3 and 4 in the susceptible class. This separation is based on the nature and size of uredinia, if any, that are produced on the test plants and the presence or absence of chlorosis and necrosis associated with the infection sites. The mesothetic infection class consists of only one infection type, designated "X." This condition is defined by the presence of both susceptible and resistant infection classes on the same leaf following inoculation with genetically homogenous spores. By comparing the responses to those obtained with previously identified races, a particular rust sample can be identified as one previously described. Alternatively, a new race can be detected and assigned a number if a previously undescribed combination of responses occurs in the set of differential varieties. Subraces (slight variants) within a race can be identified by using additional (supplemental) differential varieties. For example, race 15B is separated from race 15 by its pathogenicity on the supplemental differential, cultivar Lee. Otherwise, the responses of races 15 and 15B are the same on the standard set of 12 differentials.

Among the shortcomings in this system is the fact that the genetic composition of the differential varieties was not known for many years. Eventually, it was discovered that some of the differential varieties possessed numerous genes for rust resistance. Consequently, races capable of attacking such varieties must, by necessity, have numerous genes for pathogenicity. The genetic composition of races of *Puccinia graminis* f. sp. *tritici* identified by this system therefore was not clearly defined. Also, the genes for resistance in the standard differential varieties were no longer being used in commercial varieties. Therefore, selection pressure existed in nature for virulence genes that could not be detected by the standard differential varieties. A better system was needed.

The first step in a new direction was provided in North Dakota by H. H. Flor (1954), who constructed a set of differential varieties of flax to identify races of *Melampsora lini,* the flax rust fungus. Flor's method was to develop a series of flax lines essentially monogenic for their differential resistance to rust infection. He developed some 25 lines of flax, each containing a single, mutually exclusive gene for rust resistance, with all lines having the common genetic background of the cultivar Bison. Thus, based on which genes for resistance in the set of differentials a rust specimen would overcome, it became possible to identify races of *M. lini* by their specific gene(s) for pathogenicity. This procedure provided the experimental basis for the classic gene-for-gene theory for genetic control of pathogenicity (see Chap. 4.3.1).

Flor's system of utilizing monogenic differential lines was subsequently applied to identification of races of *Puccinia graminis* f. sp. *tritici;* thus was developed the "new" system referred to above. For example, in the United States, 12 monogenic wheat lines are presently being used, each containing a single, mutually exclusive gene for rust resistance. Similar to Flor's system with *Melampsora lini,* races of *P. graminis* f. sp. *tritici* can be identified as to their specific gene(s) for pathogenicity. Also, different monogenic differentials can be substituted and/or added to detect specific changes in genetic composition of rust races in nature.

No longer are numbers used to designate races, as in the Stakman system, but a series of code numbers or letters is given that denotes the specific gene(s) present. The specific code varies among the Canadian, U.S., and Australian systems, but the equivalents are easily determined. This is a very workable system from the viewpoint of a plant breeder, who must know exactly what genes for pathogenicity exist in the natural rust population and consequently what genes for resistance are required in a cultivar to be released for commercial production. Such information greatly increases the breeder's lead time in development of new, rust-resistant cultivars.

Although monogenic differentials are used to identify races of flax, wheat stem, and wheat leaf rusts, many systems of race identification based on differential cultivars are still in use. In such systems, monogenic differentials have not yet been developed, e.g., coffee, maize, spearmint, and sunflower rusts.

The genetic plasticity of physiologic races of *Puccinia graminis* f. sp. *tritici* and the resulting incidence of stem rust on previously resistant cultivars are discussed briefly in Chapters 1.4 and 7.4. Graphic evidence of this is seen in Figure 3.5, based on individual races or closely related race groups in Canada during the period 1919–1955. Similarly, such shifts in population can occur within races, e.g., race 15B (Fig. 3.6). Race 15B, a subrace of race 15, and subsequently detected variants with 15B were originally identified by the Stakman system. Later, some of

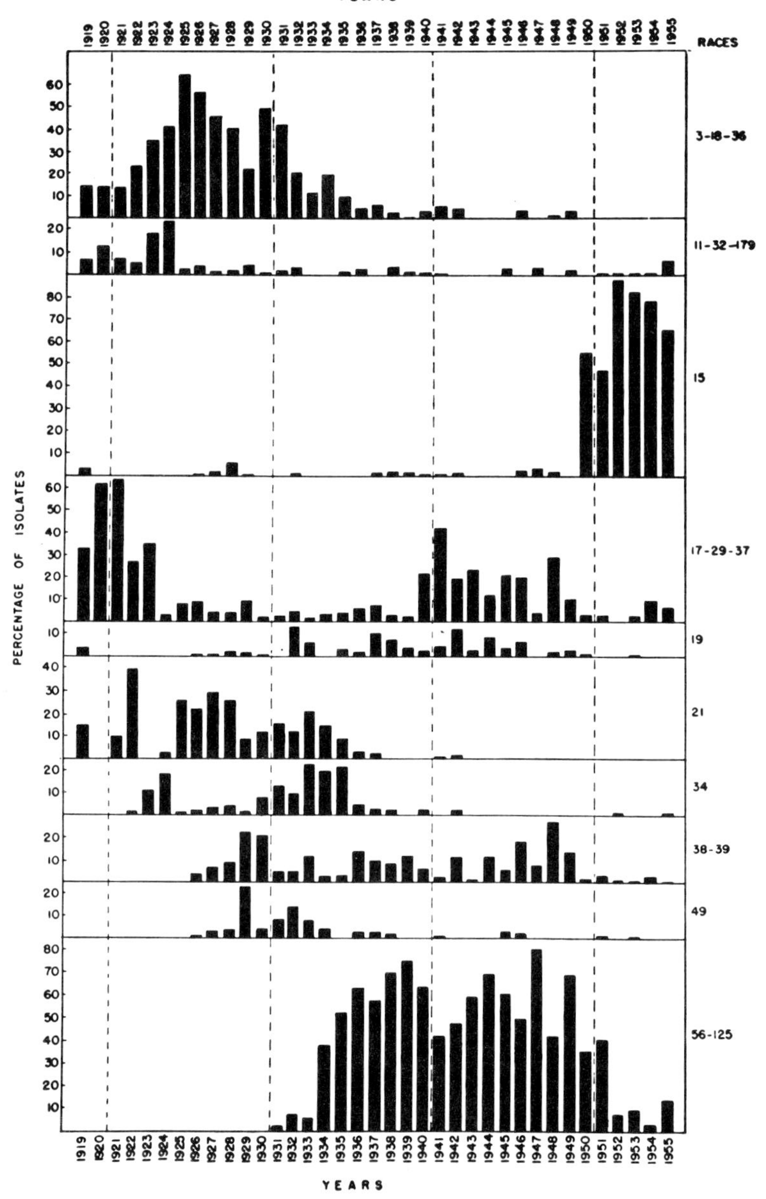

Fig. 3.5. Prevalence in Canada of 10 physiologic races or race groups of *Puccinia graminis* f. sp. *tritici* from 1919 to 1955. (From Johnson and Green 1957, by permission of the National Research Council of Canada)

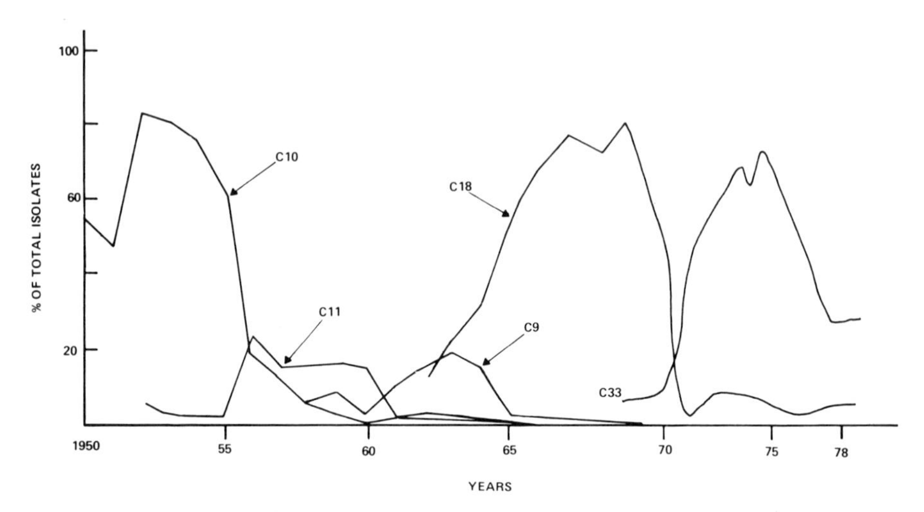

Fig. 3.6. Prevalence in Canada of 5 subraces of race 15B (*Puccinia graminis* f. sp. *tritici*) from 1950 to 1979. (Adapted from Green 1975)

them were designated in the Canadian system as races C9, C10, C11, C18, and C33, as shown in Figure 3.6. Although race C10 (15B) was first detected in 1939 in the United States, it did not become predominant until 1950, and then, suddenly. By 1952 it comprised over 80% of all stem rust isolates identified in Canada. This marked increase resulted from the ability of race C10 to attack all commercial durum and bread wheats grown in North America. Many of those cultivars had resistance bred in to combat race 56, the race responsible for the epidemic of 1935 and the demise of cultivar Ceres. After introduction of the race C10-resistant Selkirk in 1954, the prevalence of race C10 (15B) dropped sharply and was rarely found after 1960. Not to be outdone, the pathogen continued to evolve, spawning new physiologic races. Races C11 and C9 soon followed, but neither became predominant. Newer variants of race C10, i.e., races C18 and C33, became the predominant races collected, and each later declined in prevalence. Although their prevalence was high in terms of percentage of races collected, their total numbers were not correspondingly high, and neither forced the large-scale replacement of any wheat cultivar. In fact, no major stem rust epidemic has occurrred in the spring wheat regions of North America since the one resulting from race C10 (15B) in the early 1950s. However, what lies ahead in this genetic shell game is yet to be revealed.

REFERENCES

Alexopoulos, C. J., and Mims, C. W. 1979. Introductory Mycology, 3rd ed. New York, London, Sydney: John Wiley & Sons.

Brown, M. F., and Brotzman, H. G. 1979. Phytopathogenic Fungi: A Scanning Electron Microscopic Survey. Columbia: Univ. of Missouri Extension Division.

Flor, H. H. 1954. Identification of races of flax rust by lines with single rust conditioning genes. U.S. Dep. Agric. Tech. Bull. 1087.

Green, G. J. 1975. Virulence changes in *Puccinia graminis* f. sp. *tritici* in Canada. Can. J. Bot. 53:1377–1386.

Johnson, T., and Green, G. J. 1957. Physiological specialization of wheat stem rust in Canada, 1919 to 1955. Can. J. Plant Sci. 37:275–287.

Littlefield, L. J., and Heath, M. C. 1979. Ultrastructure of Rust Fungi. New York, London: Academic Press.

FURTHER READING

Cummins, G. B. 1959. Illustrated Genera of Rust Fungi. Minneapolis: Burgess.

Stakman, E. C., and Harrar, J. G. 1957. Principles of Plant Pathology. New York: Ronald Press.

Webster, J. 1980. Introduction to Fungi, 2nd ed. London, New York, Melbourne: Cambridge Univ. Press.

4

Host-Pathogen Relations

4.1 Anatomy

4.1.1 ROUTES OF PENETRATION. Different spore stages typically have their characteristic route of entry into the plant, although exceptions exist. Germinating basidiospores usually produce a short germ tube closely appressed to the host surface at the end of which is produced an appressorium (Fig. 4.1). The enlarged appressorium gives rise from its lower surface a small penetration peg that pierces the host cuticle and epidermal cell wall. The penetration peg continues growing, giving rise to mycelium within the penetrated epidermal cell. Subsequent growth is described in the next section (4.1.2). Although direct epidermal penetration is typical of germinating basidiospores, penetration through stomata occurs in some rusts, e.g., *Cronartium ribicola* on pine needles.

In contrast to basidiospores, the germlings of aeciospores and urediniospores typically penetrate through stomata after formation of germ tubes and appressoria (Fig. 4.2). In most rusts studied, the triggering of appressorium formation over the host stoma appears to be in response to some physical or thigmotropic stimulus, e.g., the raised margin of the stomatal lips. In some studies of wheat stem rust, the triggering appears to result from certain nonvolatile substances in the leaf cuticle and volatile substances emanating from within the leaf. In some tropical and semitropical rusts, e.g., soybean rust (*Phakopsora pachyrhizi*), urediniospore germlings penetrate directly through the epidermis. Teliospores, of course, do not penetrate plants upon germination but rather produce basidiospores. Likewise, pycniospores do not penetrate the host but function as spermatia in sexual reproduction (see Chap. 2.1).

4.1.2 INTER- AND INTRACELLULAR STRUCTURES. Once penetration is completed, a variety of fungal structures can be formed within

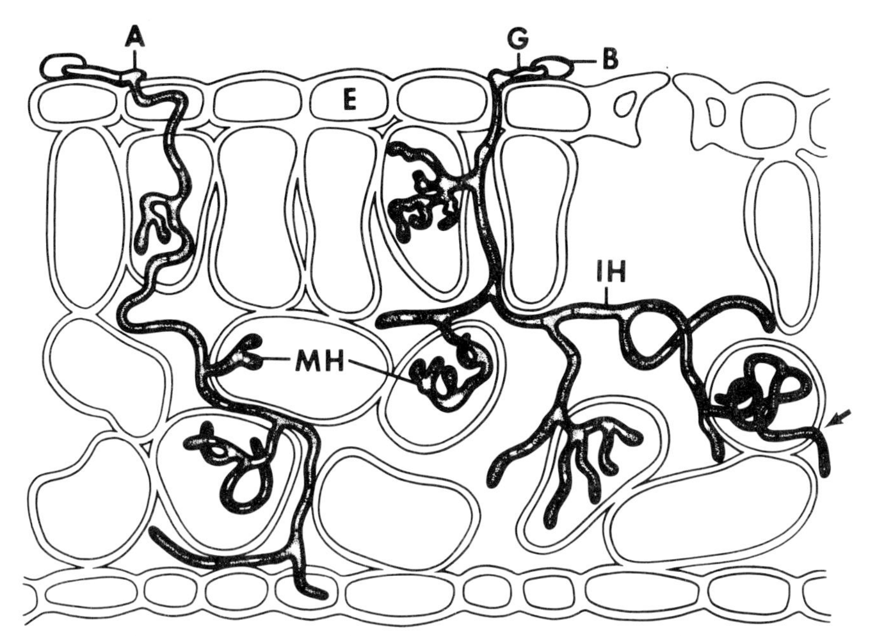

Fig. 4.1. Diagrammatic representation of a haploid stage rust infection initiated by basidiospore germlings. Basidiospore (B) germinates to produce germ tube (G), with a terminal appressorium (A). Penetration can occur through epidermal cells (left) or between them (right). Subsequent growth produces intracellular hyphae in underlying mesophyll cell (left) or intercellular hyphae (IH) (right). Monokaryotic haustoria (MH) form as branches from the intercellular hyphae and often assume a hyphalike appearance within the invaded host cell; they sometimes grow out of the cell (arrow) again, becoming intercellular in habit.

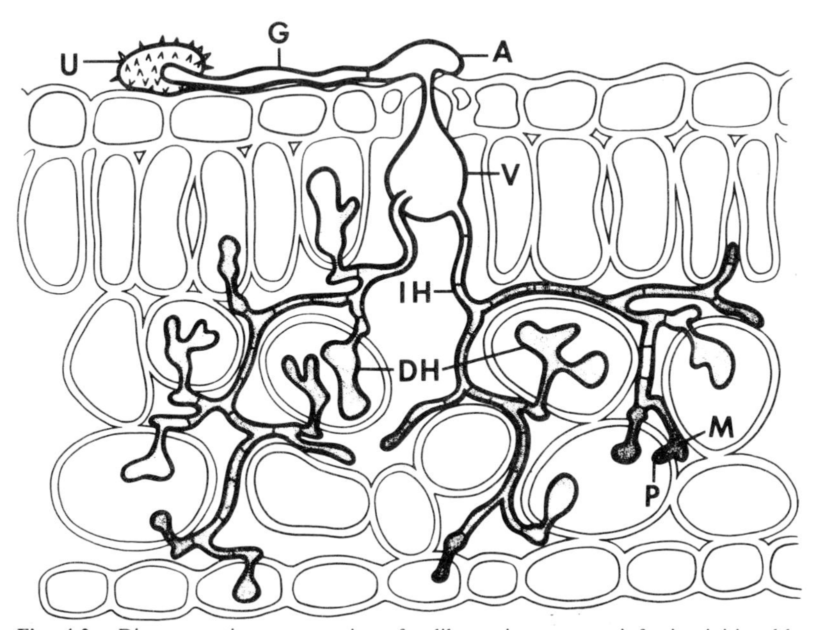

Fig. 4.2. Diagrammatic representation of a dikaryotic stage rust infection initiated by aeciospore or urediniospore germlings. The urediniospore (U) germinates to produce a germ tube (G), with terminal appressorium (A) over a host stoma. Penetration occurs through the stoma; a substomatal vesicle (V) is formed. The latter gives rise to intercellular hyphae (IH), which ramify the host tissue. Specialized terminal cells of the intercellular hyphae, i.e., the haustorial mother cells (M), produce penetration pegs (P) that enter the host cells and eventually form the spherical to lobed dikaryotic haustoria (DH).

the host tissue. The specific structures depend partly upon the stage of the life cycle, i.e., whether the developing thallus is derived from a haploid basidiospore, a dikaryotic aeciospore, or a urediniospore.

Following penetration of epidermal cells by basidiospore-derived mycelium, the 1N mycelium will grow, i.e., exit, from the epidermal cell and establish mycelium in the intercellular spaces of the host (Fig. 4.1). This mycelium can branch laterally, penetrating the mesophyll cells. The branches (Fig. 4.3) that extend into those host cells are similar morphologically to the intercellular hyphae. They are variously termed intracellular hyphae, haustoria, or, more precisely, monokaryotic haustoria (Figs. 4.1, 4.3). These structures are different morphologically from haustoria found in the dikaryotic stages of the life cycle, although they probably have similar functions. Monokaryotic haustoria are often sep-

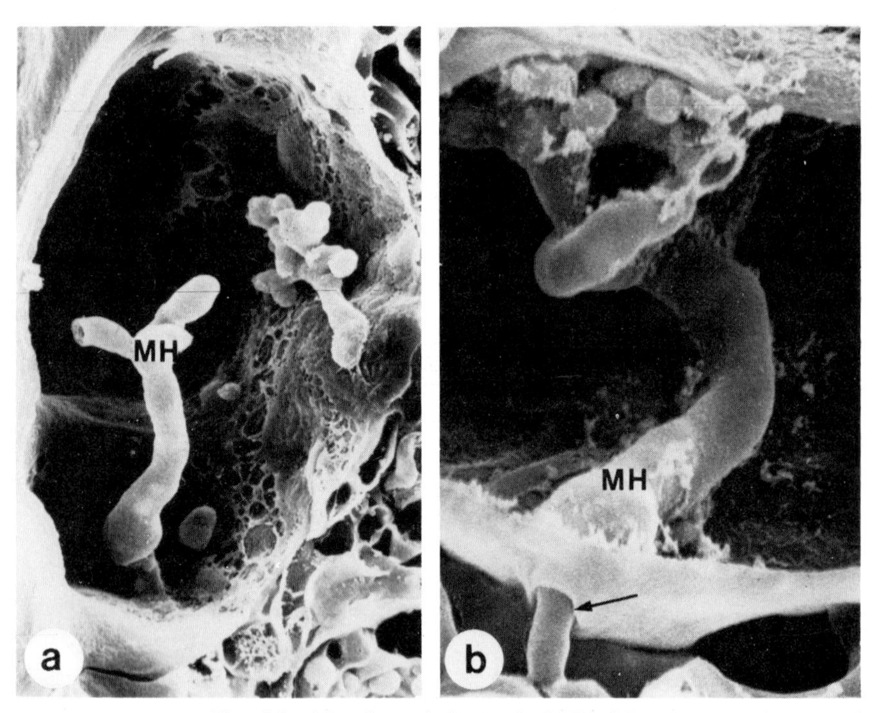

Fig. 4.3. Monokaryotic haustoria (MH) of *Puccinia recondita* (a) and *P. malvacearum* (b). Note the branched, hyphalike nature of each and the apparent growth of the latter from the host cell (arrow) to become once again intercellular. ×1900 and ×5400 respectively. (*P. recondita* from Gold et al. 1979, by permission of the National Research Council of Canada; *P. malvacearum* courtesy R. E. Gold, from Littlefield and Heath 1979, by permission of Academic Press)

tate. They may colonize the host cell extensively to form coils or masses of hyphae, and sometimes they may grow from the cell, again becoming intercellular, or into adjacent cells. The entire monokaryotic haustorium is surrounded by the invaginated cell membrane (plasmalemma) of the host cell; thus it is not in direct contact with host cytoplasm.

Morphologically, the structure of the dikaryotic thallus (Fig. 4.2) is more complex than that of the monokaryotic. Once the penetration peg has grown through the stoma, it enlarges to form a substomatal vesicle. This structure is more or less oval to spherical in shape and occupies much of the volume of the substomatal cavity. From it arise one or more infection hyphae, which form the intercellular mycelium (Fig. 4.4).

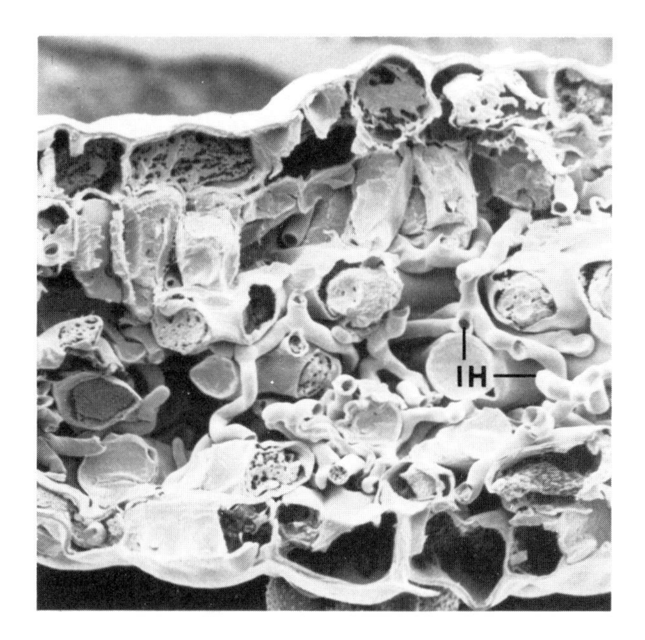

Fig. 4.4. Growth of dikaryotic intercellular hyphae (IH) of *Hemileia vastatrix* through mesophyll of coffee leaf. ×2000. (Courtesy J. Harr, Sandoz, Ltd., Basle, Switzerland, and R. Guggenheim, University of Basle)

These intercellular hyphae form small terminal branches that form haustorial mother cells. The latter structures are closely appressed to host mesophyll cells or the sides of epidermal cells toward the interior of the plant. The haustorial mother cell gives rise to a small-diameter penetration peg, which penetrates the host cell wall and proceeds to invaginate the host plasmalemma. After growing some 3–4 μm into the host cell, the apex of the penetration peg begins to expand (Fig. 4.5). This expansion eventually forms the body of the haustorium, which may be spherical to variously lobed or branched.

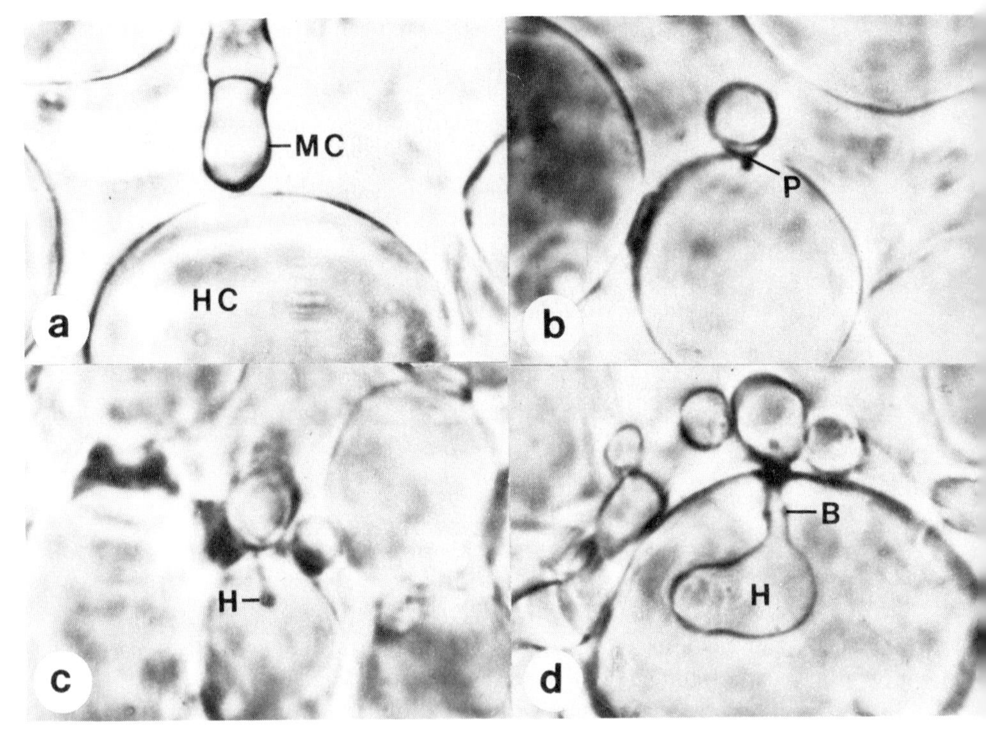

Fig. 4.5. Steps in the formation of dikaryotic haustoria of *Melampsora lini.* (a) Initial contact of haustorial mother cell (MC) with host mesophyll cell (HC). (b) Extension of penetration peg (P) into host cell. The hypha that gives rise to the haustorial mother cell is out of the plane of focus. (c) Early stage in expansion of the haustorial body (H) at the apex of the penetration peg, the latter having reached its ultimate length. (d) Mature, lobed haustorium (H) with dark-staining neckband (B) midway along haustorial neck. ×2100. (From Littlefield 1972, by permission of the National Research Council of Canada)

Dikaryotic haustoria are not septate, and they are borne on a narrow necklike structure (Figs. 4.5, 4.6) that previously had functioned as the penetration peg. Dikaryotic haustoria are always terminal structures, never growing from the host cell. Haustoria, both mono- and dikaryotic, are organs that probably function in absorption of nutrients from the host. As with monokaryotic haustoria, the dikaryotic haustoria reside outside the invaginated host plasma membrane. A material termed the "extrahaustorial matrix" occupies the space between the haustorial cell wall and the invaginated host plasma membrane. The composition of this matrix is uncertain.

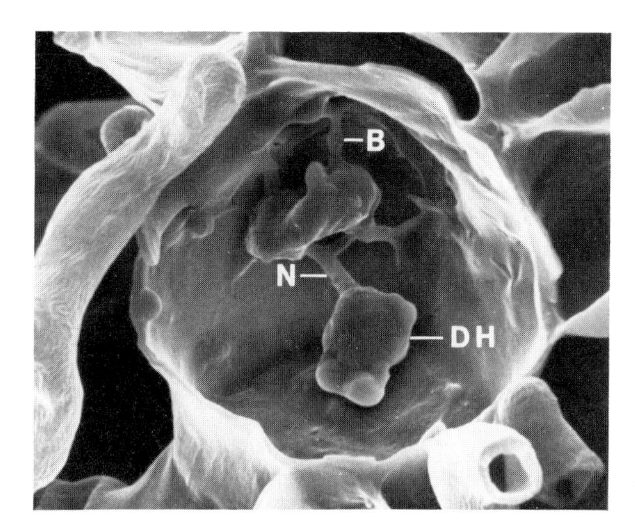

Fig. 4.6. Dikaryotic haustoria (DH) of *Hemileia vastatrix* in mesophyll cell of coffee leaf. Three haustoria appear to be present in the cell. Note the elongated haustorial neck (N) and the enlarged neckband (B) typical of dikaryotic haustoria. ×2400. (Courtesy J. Harr, Sandoz, Ltd., Basle, Switzerland, and R. Guggenheim, University of Basle)

4.1.3 ABNORMAL GROWTH RESPONSES. Rust infections may induce numerous deviations from normal growth patterns, especially when infection occurs systemically or in meristematic tissues. Such changes include dwarfing, the shift from prostrate to erect growth habit, and various tissue malformations.

Dwarfing (hypoplasia) is generally an indirect effect and typically is not intense. Smaller leaves and fruits, shorter stems, and general stunting may result from the diversion of nutrients from host to parasite or from the inhibition of their synthesis. Severe stunting of snapdragon can occur, however, when infected at an early age by *Puccinia antirrhini.* Not only is the plant reduced in size, but leaves become thicker and shorter and inflorescences partially to totally abort.

A change from the normally prostrate to erect growth occurs with systemic infection of the leafy groundwort (*Hepatica acutiloba*) by *Tranzschelia punctata.*

Tissue malformation, if it occurs, is usually but not always associated with sorus formation, especially aecia (Fig. 4.7). Such malformations include hypertrophy, the enlargement of a tissue brought about by the massing of hyphae between host cells and the expansion of neighboring parenchyma cells. Also associated with hypertrophy are increased levels of the growth hormones auxin and gibberellin in the enlarged host organs. The degree of distortion depends partly on the extent of infection and the coalescence of infected areas, especially when associated with active meristematic tissue.

Fig. 4.7. Enlarged, twisted stem of clematis with aecial gall (arrow) produced by *Puccinia recondita.* The gall surface is covered with the cup-shaped aecidioid aecia typical of the genus *Puccinia.*

FIG. 4.8. Multiple infection of lodgepole pine by *Endocronartium harknessii* resulting in numerous oblong to pear-shaped galls. (From Ziller 1974, reproduced by permission of the Minister of Supply and Services Canada)

Hypertrophy of leaves often causes their localized thickening and buckling, forming raised areas on which pycnia and aecia develop (see Fig. 2.8). Hypertrophy of stems often causes formation of galls. Galls develop when there is marked overgrowth in an organ, especially when the swelling is circumscribed and more or less abrupt. Galls may be globose, lobed, or fusiform (spindle shaped); they may be annual or perennial; and they may occur on herbaceous or woody tissue (Fig. 4.8). Galls on woody tissues are common in *Cronartium* spp. (e.g., pycnial/aecial stages on pine), *Gymnosporangium* spp. (telial stage on junipers, see Plate 3.1), and *Uromycladium* spp. (telial stage on acacia, e.g., in Australia, Fig. 4.9).

Fig. 4.9. Elongated telial galls on branches of acacia caused by *Uromycladium tepperianum*. (Courtesy J. Walker, Australia Department of Agriculture)

Another type of tissue malformation is the so-called witches'-broom (Figs. 4.10–4.12) found on certain woody shrubs and trees. These result from infection of growing points and the subsequent stimulation of excessive bud formation. The buds elongate to produce the compact, bushy witches'-broom, some of which can persist 15–20 years. The mycelium can invade the woody tissue at the base of the broom and survive there in a vegetative state for many decades.

Fig. 4.10. Top of a mature alpine fir killed by aecial stage witches'-broom (arrow) caused by *Melampsorella caryophyllacearum*. (See Figure 2.10 for close-up view of aecial sori and the distorted host needles in the witches'-broom.) (From Ziller 1974, reproduced by permission of the Minister of Supply and Services Canada)

Fig. 4.11. Telial stage witches'-broom on evergreen huckleberry caused by *Pucciniastrum geoppertianum*. (From Ziller 1974, reproduced by permission of the Minister of Supply and Services Canada)

In perennial witches'-brooms, e.g., *Melampsorella* sp. on fir (Fig. 4.10) and *Chrysomyxa* sp. on spruce, the host needles bear pycnia and aecia and are shed yearly. Such needles are shorter and broader and have an inflated appearance compared to healthy needles (see Fig. 2.10). Perennial witches'-brooms are devoid of or greatly reduced in chlorophyll content and place a considerable nutrient drain on the host. They are important economically in the Rocky Mountains of the United States, where they cause reduced growth increment, death of branches and main stems distal to the broom, and mortality. Witches'-brooms caused by telial stage mycelium, e.g., *Gymnosporangium nidus-avis* on juniper (Fig. 4.12), are not chlorotic, nor do they cause annual defoliation as do those associated with the pycnial/aecial stages. Also, in the former, the symptom of rejuvenescence is evident. Needle morphology within the broom, even on mature branches, reverts to that of the juvenile stage. The rather flattened, more blunt-tipped needles of mature character are replaced by the circular, sharp-pointed needles typical of immature branches of *Juniperis*.

Fig. 4.12. Telial stage witches'-broom (arrow) on Rocky Mountain juniper caused by *Gymnosporangium nidus-avis*. (From Ziller 1974, reproduced by permission of the Minister of Supply and Services Canada)

4.2 Physiology

The physiologic basis of host-pathogen interactions in rusts, as in any disease, is highly complex. Since the numerous physiologic processes of the host—respiration, photosynthesis, translocation, and hormonal controls—are all interrelated, it is difficult to discuss any one process separately. Bearing this in mind, the following sections summarize briefly the major effects rust infections have on plants. To a great extent these effects are all related to carbon balance in plants and its imbalance brought on by infection. Large pools of organic reserves are required for spore formation and their subsequent germination. Consequently, a successful parasite must create a carbon economy in the host favorable to itself. In some instances these induced physiologic changes extend into tissues, e.g., roots, far beyond the sites of infection.

4.2.1 RESPIRATION. Respiration is the oxidation of organic materials to simpler compounds. Respiration may be anaerobic, occurring in the absence of atmospheric oxygen, e.g., fermentation; it may also be aerobic, requiring the uptake of oxygen from the air. Typically, in plants both types of respiration occur simultaneously. Figure 4.13 illustrates some of the more common pathways by which respiration occurs.

A major difficulty in studying respiration in diseased plants is determining what contribution is made by the host and what is made by the pathogen.

One cffect of rust infection is a significant increase in the rate of respiration, as determined by O_2 consumption. Upon sporulation of the

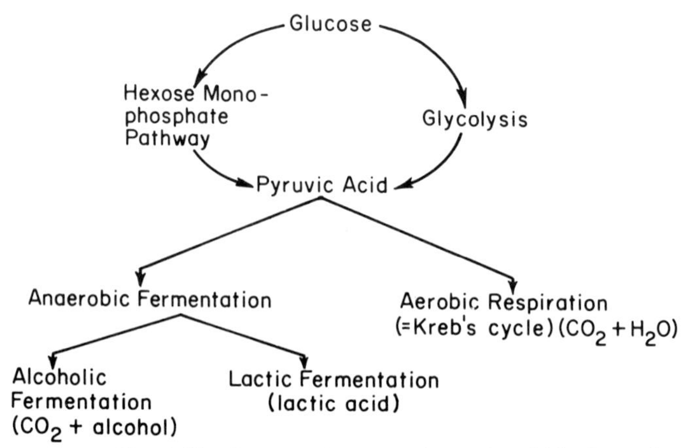

Fig. 4.13. Some respiratory pathways by which glucose is oxidized.

rust, usually 6–8 days after inoculation, the respiration rate is often 100–200% that of healthy plants. Since this is a time of intense metabolic activity of the fungus, most of this increase is thought to be from the pathogen, not the host. Furthermore, microrespirometric studies have shown that negligible increases in the respiration rate occur beyond the margin of hyphae in individual infection sites. During the time of sporulation, the fungus requires much energy in the synthesis of building blocks and storage carbohydrates that go into spore production.

Another effect of rust infection is a shift in respiratory pathways (Fig. 4.13). Typically, in rust-infected tissue, the hexose monophosphate (HMP) pathway supplants much of the initial catabolism of glucose that commonly occurs via glycolysis. This was shown by several investigators using glucose specifically labeled with ^{14}C in either the C-6 or C-1 position of the molecule. When glucose is catabolized solely via glycolysis, the release of C-6 and C-1 carbon is equally rapid; thus the C-6:C-1 ratio in the residual carbohydrate is 1. If the HMP is the only route of glucose metabolism, the C-1 is first evolved as CO_2. By monitoring the decline in the C-6:C-1 ratio, one can document the shift from glycolysis to HMP metabolism. In healthy tissue the C-6:C-1 ratio is about 0.6, whereas in rusted tissue it is about 0.2. Again, however, it is difficult to ascertain whether this shift represents changes in host cell metabolism or changes in the increasingly predominant fungus cells as sporulation approaches.

4.2.2 PHOTOSYNTHESIS AND TRANSLOCATION. Rust infection diminishes photosynthesis as a result, partially, of reduced chlorophyll content in diseased tissue. Although the photosynthesis rate remains unchanged or is even enhanced somewhat during early stages of infection, it is eventually reduced as host tissue becomes chlorotic and senesces prematurely. Also, in some rust infections, small localized areas around rust pustules retain their chlorophyll, becoming "green islands" in the generally chlorotic leaf tissue. These green islands continue photosynthesis, providing a local source of nutrients. The enhanced photosynthesis around young infection sites increases starch synthesis and its localized storage. In darkness the starch is hydrolyzed and its carbon is incorporated by the fungus, as demonstrated by the use of $^{14}CO_2$ labeling.

Although photosynthesis is eventually reduced, infected leaves often increase in dry weight more rapidly than do neighboring healthy leaves, and they retain a higher proportion of their photosynthate. Simultaneously, infection induces directed transport of materials, including carbohydrates and minerals, from uninfected tissues into infection sites. Such localized areas of metabolite concentration are referred to as "metabolic sinks." They result perhaps as an effect of cytokinen, a growth hormone, produced in many diseased plants, including rusts. Also, the fixation of CO_2 in uninfected leaves of bean can be enhanced by rust infection in neighboring leaves, thus providing more total carbohydrate for export into the diseased leaves. In some instances, e.g., gall formation associated with aecium development, increased dry weight per unit volume of tissue is accompanied by increased size of host cells, i.e., hypertrophy. Thus numerous factors combine to provide a net increase in the amount of dry matter present in rusted tissue.

It is almost impossible to assess what proportion of this increased dry matter is present in host cells and how much represents that in the fungus colony. That a significant portion of concentrated materials reside in the host, not the pathogen, is indicated in bean rust. There, selective killing of the pathogen by heat treatment after infection does not fully diminish the directed transport of materials into the infection sites.

The diversion of nutrients from healthy into diseased tissues reduces growth of roots, fruits, and other organs, contributing to yield loss in rusted plants.

4.2.3 WATER BALANCE. Rust infection has marked effects directly and indirectly on water balance in the infected plant. In early stages of bean rust infection, transpiration is reduced because of stomatal closure caused by the infection. However, once the epidermis and cuticle are ruptured (see Plate 2.2; Fig. 2.15) by developing sori, the transpiration rate immediately rises and continues at a high level, causing extreme predisposition to drought injury. Indirectly, rust infections can reduce root growth, thus leading to water stress symptoms as the soil dries out.

4.2.4 TRANSFER OF METABOLITES FROM HOST TO PARASITE. The directed translocation of nutrients within the plant to infection sites provides a ready supply of metabolites for the rust pathogen. What proportion of those metabolites occurs in the apoplastic "free space" of the tissue, e.g., intercellular spaces and freely permeable cell walls, versus that within host cytoplasm is unknown. Metabolites in the latter are separated from the pathogen by the semipermeable plasma membrane, i.e., cell membrane, of the host cells. However, in several diseases, rusts included, infection stimulates leakage of electrolytes from the host through the plasma membrane, presumably increasing their availability to the pathogen. Some of the metabolites absorbed by the fungus are utilized by hyphae in the vicinity where they were absorbed, but most of them are translocated through the hyphae to areas of spore production. There they are incorporated into organelles, cell walls, and storage products.

The question of nutrient exchange has received limited study, and most of our knowledge is based on carbohydrate exchange. A system that has been studied intensively is *Puccinia poarum* on coltsfoot (pycnial/aecial stages) and meadow grass (uredinial/telial stages).

The major product of photosynthesis, and that translocated throughout the plant, is sucrose. Sucrose most probably must be broken into its component hexoses—glucose and fructose—by the essential enzyme invertase before it can be absorbed by the rust fungus. Increased levels of invertase occur in rusted coltsfoot. The hexoses produced by inversion of sucrose can then be taken up by the fungus, where it is converted into trehalose (a common storage form of sugar in fungi) and various sugar alcohols (polyols), e.g., mannitol and arabitol. These compounds cannot be utilized by the host plant. Thus, in effect, this sequestering of carbon as carbohydrates and lipids by the fungus may help to maintain a concentration gradient and so insure a continued flow from the host.

A summary of this system is diagrammed in Figure 4.14. Actually,

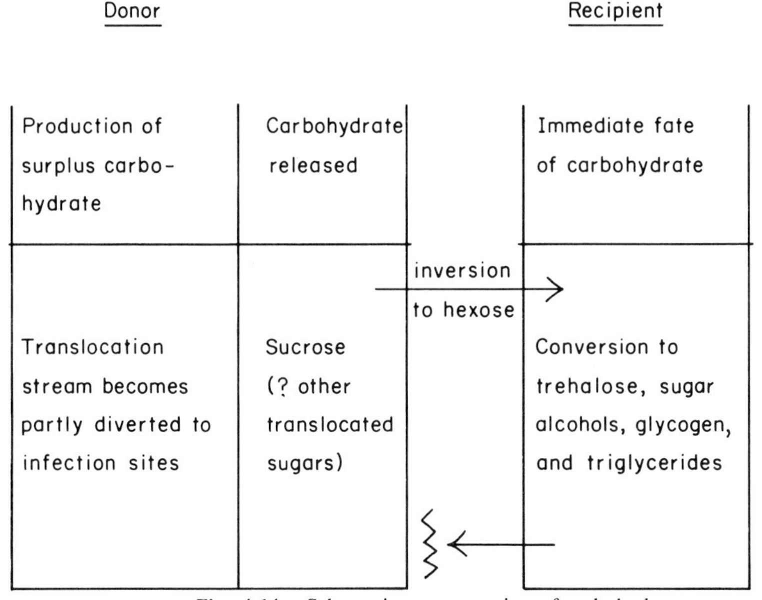

Fig. 4.14. Schematic representation of carbohydrate movement from host to rust parasite. Host cannot utilize trehalose and sugar alcohols as indicated by blocked arrow. (Adapted from Smith et al. 1969, by permission of Cambridge University Press)

for several reasons, the situation is considerably more complex than that shown in the diagram. Presumably, the fungus can absorb carbohydrate through both the intercellular hyphae and the intracellular haustoria. Also, chloroplasts are usually destroyed in the central areas of lesions, although they remain functional for a time in the peripheral cells of the infection site even if haustoria are present. Thus starch formation and accumulation occur around the rust pustules, not within them. However, fructans can accumulate in host cells within the pustule area. A more precise diagram of carbohydrate exchange is shown in Figure 4.15.

In coltsfoot rust several amino acids as well as carbohydrates concentrate in the infection sites. Some amino acids, e.g., serine and alanine, are absorbed from the host by the fungus more readily than others, e.g., glutamine and glutamic and aspartic acids. It has been suggested, therefore, that certain host-derived amino acids are more important than others in the successful establishment and maintenance of the rust on the host.

Similarly, radio-labeled pyrimidine nucleosides, but not purine nucleosides, are taken up from the host by the oat crown rust fungus. These components of nucleic acids are apparently absorbed only through haustoria since they are not taken up by the early-stage intercellular substomatal vesicles and infection hyphae as is ^{3}H-glucose.

In wheat stem rust infection, hexoses absorbed from the host are synthesized partly into fungal cell wall material or fatty acids. The fungus does not absorb intact proteins from the host, rather, free amino acids, which are synthesized into proteins by the fungus. Some fungal amino acids, e.g., 70% of the spore-alanine, are synthesized by the fungus itself, not absorbed directly from the host.

4.3 Incompatibility

Compatibility and incompatibility are corporate responses of the host and the pathogen, both organisms being intimately associated in an antagonistic type of symbiosis. Macroscopic symptoms of the disease as well as anatomic and biochemical reactions that lead to the expression or to the lack of symptoms depend upon both organisms. Thus one must consider compatibility or incompatibility of particular host-pathogen combinations, not simply whether a host is resistant or susceptible. A particular host may be either resistant or susceptible, depending on its own genetic potential, the genetic potential of the pathogen, and even the environment.

In this section, we will consider various genetic, morphologic, and

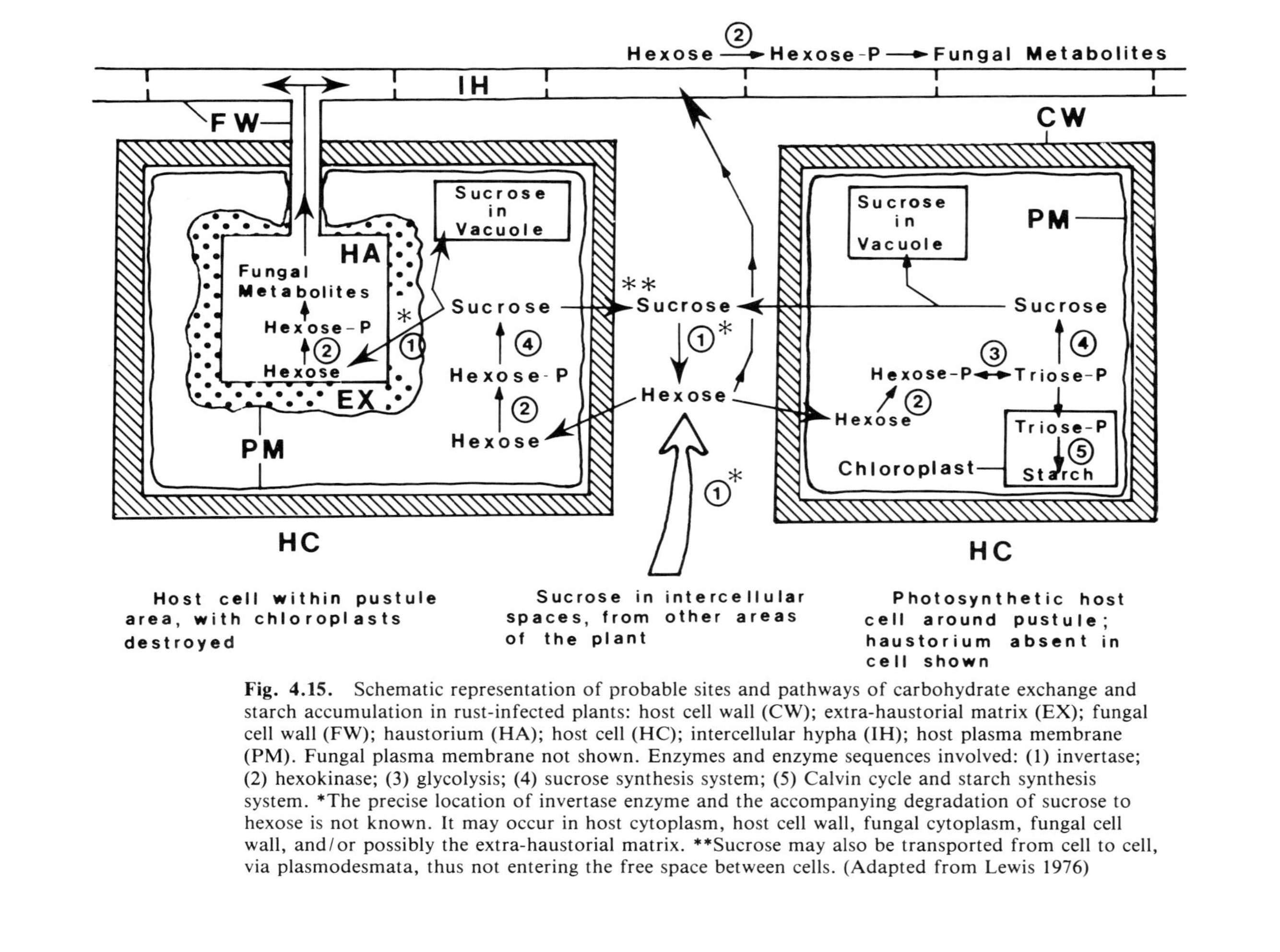

Fig. 4.15. Schematic representation of probable sites and pathways of carbohydrate exchange and starch accumulation in rust-infected plants: host cell wall (CW); extra-haustorial matrix (EX); fungal cell wall (FW); haustorium (HA); host cell (HC); intercellular hypha (IH); host plasma membrane (PM). Fungal plasma membrane not shown. Enzymes and enzyme sequences involved: (1) invertase; (2) hexokinase; (3) glycolysis; (4) sucrose synthesis system; (5) Calvin cycle and starch synthesis system. *The precise location of invertase enzyme and the accompanying degradation of sucrose to hexose is not known. It may occur in host cytoplasm, host cell wall, fungal cytoplasm, fungal cell wall, and/or possibly the extra-haustorial matrix. **Sucrose may also be transported from cell to cell, via plasmodesmata, thus not entering the free space between cells. (Adapted from Lewis 1976)

biochemical interactions between host and pathogen that result in incompatible interactions. As in the previous discussion of disease physiology, it is impossible to consider incompatibility in terms of individual genetic, biochemical, and morphologic components. Morphologic alterations are the results of biochemical processes, which in turn are controlled by genetic potentials of the two symbionts.

4.3.1 GENETIC CONTROLS. Since the determination of compatibility or incompatibility is under genetic control of both host and pathogen, it is impossible to consider a host cultivar as resistant or susceptible without specifying the genotype of the pathogen. This control by specific complementary genes in the host for resistance or susceptibility and for virulence and avirulence in the pathogen is the basis for Flor's (1956) gene-for-gene theory. With few exceptions, resistance to rust is dominant in the host, and avirulence is dominant in the pathogen. The nature of this relationship is shown, simplified, in Figure 4.16. The host with either the RR or Rr resistance genotype is resistant only to the AA or Aa avirulence genotype of the pathogen. If the RR or Rr genotype of the host is inoculated with the aa (virulent) genotype, a compatible reaction will result. In the absence of R, the host is susceptible, irrespective if the pathogen is A– or aa. This type of relationship occurs in other plant diseases, e.g., powdery mildews, as well as rust diseases.

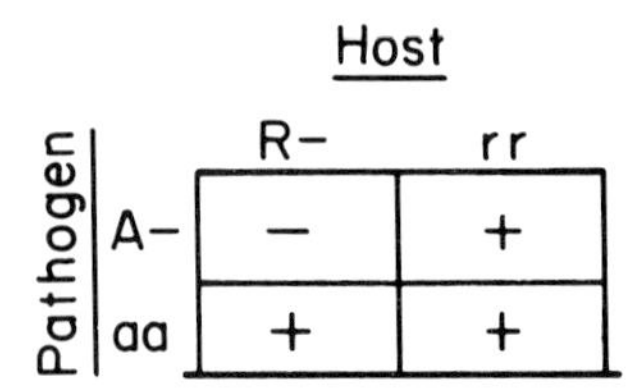

Fig. 4.16. The pattern of interaction between host and pathogen in most rust diseases: dominant avirulence in the pathogen (A); dominant resistance in the host (R), compatible and incompatible reactions, (+) and (–) respectively.

Specifically, Flor demonstrated that in flax rust five different host genes (designated K, L, M, N, and P) can confer resistance when they occur in the dominant condition. Those dominant host genes can confer resistance, however, *only* when the race of the pathogen contains the corresponding dominant genes for *a*virulence (thus the name, "gene-for-gene" relationship). Flor designated those corresponding genes for avirulence in the pathogen as A_k, A_l, A_m, A_n, and A_p and the recessive genes for virulence as a_ka_k, a_la_l, a_ma_m, a_na_n, and a_pa_p. However, in order to simplify the discussion, the genes in both the host and pathogen are referred to here simply as K, L, M, N, P (dominant) or k, l, m, n, p (recessive) and as being clearly identified as belonging to either the host or the pathogen. Also, as an additional complication, in the host are multiple alleles at each of the gene loci except K; thus K, L . . . L_{12}, M . . . M_6, N . . . N_2, and P . . . P_4 comprise a total of 29 alleles identified to date that govern the interaction. There are likewise 29 multiple alleles for avirulence/virulence in the pathogen. However, since the multiple alleles in the host are thought to be mutually exclusive, it is presumably impossible to synthesize a flax host with more than 5 total alleles for resistance.

The gene-for-gene relationship in flax rust has been illustrated as the "lock and key" system by Browning (1963), as shown in Figure 4.17. If the host has no resistance "locks" (i.e., assuming resistance is dominant, it has recessive genes at each locus that conditions disease reaction), then no particular "key" is needed to gain entrance, and the reaction is one of susceptibility (Fig. 4.17, line 1). If, however, the host has "lock" K_ and the pathogen has no "key," then "lock" K_ effectively limits the advance of the pathogen, and the reaction is one of resistance (line 2). On the other hand, if the pathogen possesses "key" kk specific for "lock" K_, the reaction is susceptible (line 3). The addition of "lock" L_ renders the host resistant to the pathogen with only gene kk (line 4) or gene ll (line 5), but the pathogen with "keys" kk and ll is able to attack successfully a host with "locks" K_ and L_ (line 6). The addition of a third "lock," M_, excludes the pathogen with only "keys" kk and ll, however, as M_ becomes the limiting factor in the host: parasite interaction (line 7). The addition of "key" mm (line 8) enables the pathogen to attack host K_L_M_. Additional "keys," nn and pp, neither help nor hinder the attack on K_L_M_ but may be carried in reserve until needed on a host with some combination of "locks" K_, L_, M_, N_, and P_. A host plant with these five "locks" is resistant to any race of the pathogen that is deficient in one or more of the five necessary pathogenicity "keys." Thus the absence of "key" kk for "lock" K_ becomes the limiting factor in the development of pathogen K_llmmnnpp on host K_L_M_N_P_ (line 9).

An example of such genetic control of flax rust is the cultivar Bison, which contains the dominant L_9 gene for resistance. It can be attacked only by those races containing the homozygous recessive l_9l_9 gene for virulence. Other races containing L_9L_9 or L_9l_9, both conditions conveying avirulence, cannot attack the Bison cultivar. Such genotypes exist in Australia but not in North America. Consequently, Bison is susceptible to all North American races of *M. lini,* since the latter are all homozygous recessive for l_9l_9, which conditions virulence.

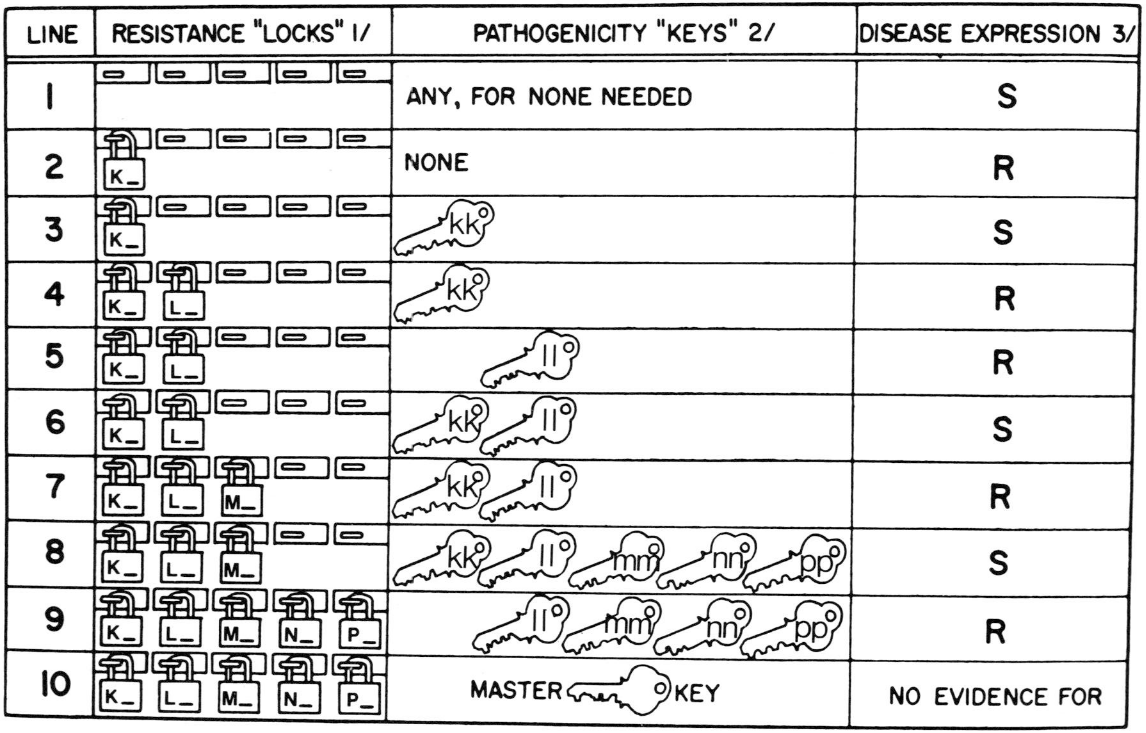

LINE	RESISTANCE "LOCKS" 1/	PATHOGENICITY "KEYS" 2/	DISEASE EXPRESSION 3/
1		ANY, FOR NONE NEEDED	S
2	K_	NONE	R
3	K_	kk	S
4	K_ L_	kk	R
5	K_ L_	ll	R
6	K_ L_	kk ll	S
7	K_ L_ M_	kk ll	R
8	K_ L_ M_	kk ll mm nn pp	S
9	K_ L_ M_ N_ P_	ll mm nn pp	R
10	K_ L_ M_ N_ P_	MASTER KEY	NO EVIDENCE FOR

Fig. 4.17. Resistance ''locks'' and pathogenicity ''keys'' used to illustrate the gene-for-gene hypothesis for interactions in host-parasite systems. (1) The ''hasp'' represents a locus at which any one of an allelic series of ''locks'' can occur. The absence of a ''lock'' on a given ''hasp'' indicates the recessive allele, and no ''effective lock'' is present. (2) The absence of a given ''key'' indicates the dominant allele, and no ''effective key'' is present. (3) S indicates susceptible, R, resistant. (Adapted from Browning 1963, by permission of the Iowa Academy of Sciences)

One interpretation of this kind of interaction is that a specific interaction for incompatibility is operative. A suggested mechanism for such a system, based on eliciting substances and subsequent responses, is presented in Figure 4.18. Such a mechanism is based on the assumptions that (1) the pathogen "triggers" a response in the host (therefore the host must "recognize" the avirulent pathogen in order to respond) and (2) the host responds to the triggered response in the pathogen and forms substances and/or undergoes responses that are injurious to the pathogen. Possible chemicals involved and cytologic effects produced by those chemicals are discussed in the following sections.

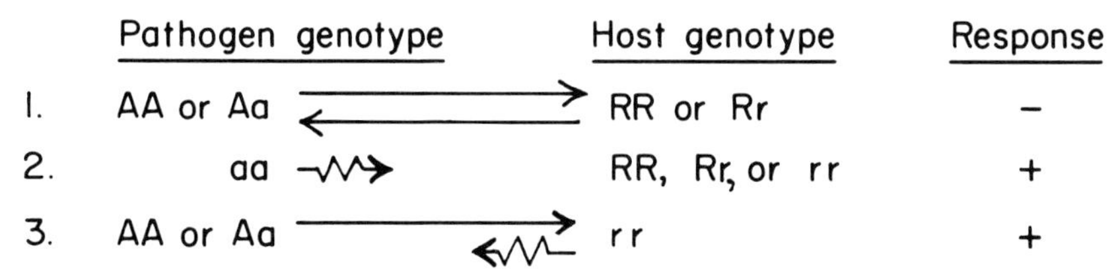

Fig. 4.18. Possible mechanism of interaction between pathogen and host in rust diseases. (–) and (+) = incompatibility and compatibility, respectively. In no. 1 the pathogen produces "triggering" substances that initiate responses in the host that are deleterious to the pathogen, indicated by the two arrows. In no. 2 the pathogen lacks the ability to elicit a response by the host; consequently the interaction is compatible. In no. 3 the pathogen produces "triggering" substances, indicated by the arrow, but the host lacks the genetic potential to respond, thus leading to compatibility. "A" indicates dominant avirulence in the pathogen; "R" indicates dominant resistance in the host.

4.3.2 BIOCHEMICAL RESPONSES. Many biochemical responses (e.g., altered metabolic pathways, stimulated syntheses, inhibited syntheses) have been associated with incompatible reactions to rust infection. Many such responses have been claimed to be responsible for the observed incompatibility. Often, though, the critical cause-effect relationship has not been thoroughly examined. Thus what may appear to be a cause of or mechanism for incompatibility may be only a symptom, i.e., a spurious event that is an outcome of a causative event, *not* the causative event itself. Also, the mistake must be avoided in thinking of *the* mechanism of rust resistance. Biochemical and anatomic studies indicate that several mechanisms of rust resistance may exist, especially when one compares highly immune and intermediate-resistant responses of a host to infection.

With rare exceptions, resistance to any plant disease, including rusts, is "triggered" or induced by the invading pathogen. Almost never is the prevention of infection due to preexisting inhibitory substances or morphologic properties of the host. The inducing principle is thought to be a diffusible substance(s) emanating from the pathogen that the host "recognizes" as foreign. This substance is referred to as an "elicitor." The elicitor then initiates the biochemical responses in the host cells that lead to incompatibility (Fig. 4.18). When such recognition and elicitation occur is not precisely known and may vary depending on the particular host-pathogen combination studied. Although documented in other diseases, elicitors and their resultant effects have not been proven in the rusts and as yet are only speculative.

One group of compounds thought to result from the action of elicitors is phytoalexins. Phytoalexins, a chemically diverse group of substances, are antibiotic compounds produced within host tissue; they are induced by microbial infection, mechanical injury, drug application, or other stimuli. They are inhibitory to pathogens at very low concentrations and display various degrees of specificity. Two such compounds, coniferyl alcohol and coniferyl aldehyde, are stimulated in certain incompatible flax rust infections (Fig. 4.19). They form more rapidly in incompatible than in compatible reactions. They are also associated with more rapid restriction of fungal growth and smaller lesion formation in highly incompatible reactions compared to less rapid restriction of the fungus and the larger lesions formed in moderately incompatible reac-

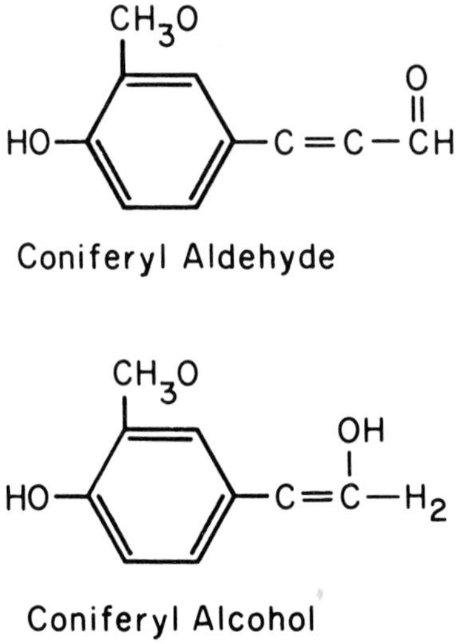

CH_3O

HO— —C=C—CH(=O)

Coniferyl Aldehyde

CH_3O

HO— —C=C(OH)—H_2

Coniferyl Alcohol

Fig. 4.19. Two compounds, coniferyl aldehyde and coniferyl alcohol, isolated from rust-infected flax that exhibit properties of phytoalexins and may function in limiting host colonization by the pathogen.

tions. However, these compounds, as well as other phytoalexins, have yet to be proven unequivocally to be the cause of the incompatible response rather than an accompanying effect of some more basic response such as cell death.

Other possible biochemical bases of incompatibility include antigenic reactions between host and pathogen and specific recognition reactions based on carbohydrate-protein bonding between the two. For example, fragments of bean rust urediniospore germ tube walls will attach to cell walls of compatible hosts. They will not, however, bind to cell walls of nonhost species in which haustoria are not formed. Limited attachment occurs onto walls of nonhosts in which a few haustoria form prior to cessation of infection. This suggests that formation of haustoria is induced by the specific attachment of the hyphal cell wall to the host cell wall, and failure to attach precludes further development of the pathogen in the host. Obviously, much research remains to be done to explain host-pathogen incompatibility.

4.3.3 MORPHOLOGIC RESPONSES. The nature and extent of morphologic responses to rust infection depend partly upon whether nonhost or cultivar resistance is being considered. A nonhost is a plant species, in its entirety, that is not susceptible to infection by any genotype of a particular rust species. Cultivar resistance is the resistance of certain cultivars of a host species to particular genotypes of a rust species.

As stated previously, morphologic properties of the plant rarely prevent rust infection. An exception, however, appears to be surface properties of wheat and oat leaves that preclude penetration by the bean rust fungus (*Uromyces phaseoli* f. sp. *typica*). There, the stomatal margins (lips) are insufficiently developed to trigger the differentiation of the appressorium, a tactile response. Thus the elongating germ tube simply grows over the stoma, not "recognizing" it as a port of entry, and eventually dies on the leaf surface.

On numerous nonhosts, however, the rust fungus will grow into the leaf, where it commonly dies prior to formation of haustoria. The fungus fails to penetrate host cells for various reasons. In some instances it results from failure of the haustorial mother cell to adhere to the host mesophyll cell wall; in others a dense wall-like deposit is formed on the inner face of the host cell wall in the region of fungal contact and presumably prevents penetration; in still others the cytoplasm of the haustorial mother cell and intercellular hyphae completely disorganizes, resulting in fungal necrosis without damage to surrounding host cells.

Cultivar resistance is exhibited by a wide range of responses.

Typically, the response in compatible and incompatible reactions is similar for the first 24–36 hours after inoculation. Visibly discernible antagonistic responses usually do not occur until after formation of the first haustorium. One of the most common observations in highly resistant (often termed "immune") hosts is the so-called hypersensitive reaction, i.e., the rapid death of the invaded cell followed by death of the pathogen and several surrounding host cells (Fig. 4.20).

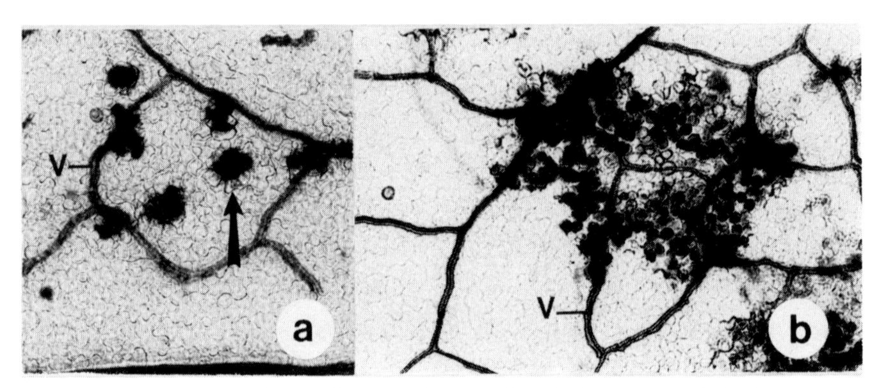

Fig. 4.20. Hypersensitive reactions in flax leaves to infection by incompatible races of *Melampsora lini.* Light microscopic view of cleared leaf tissue with necrotic areas around infection sites stained darkly. (a) Extremely rapid response limited the infection and associated necrosis (arrow) to usually less than 10 host cells. (b) Slower response allowed the fungus to colonize a greater area of host tissue (yet microscopic in size) before onset of necrosis in this single infection site; host vascular bundle (V). Both ×70. (From Littlefield 1973, by permission of Academic Press)

Not all combinations of incompatible host and pathogen genotypes, however, respond at the same rate, although a small necrotic fleck symptom ultimately results from many such combinations. The necrotic fleck with no further development of the pathogen is regarded as an immune response. Whether the different rates of development of necrosis represent different biochemical mechanisms is unknown. Such hypersensitivity has traditionally been considered a mechanism of resistance, having been so defined by Stakman in 1915. However, whether this rapid death of host cells is a cause or a consequence of incompatibility has been questioned, and further study is required for an answer.

The clearly defined morphologic responses described above are characteristic of highly specific resistance, usually conferred by one or a few genes. Other manifestations of incompatibility, often controlled by numerous additive genes (but not always), may not be associated with hypersensitivity. For example, in certain "slow-rusting" cultivars, the pathogen may develop normally but at a much slower rate than the compatible, normally "fast-rusting" cultivars. Thus a longer time is required for sori to develop, and often fewer numbers of spores per sori are produced. Both of these factors can reduce the rate of epidemic development in the field (see Chap. 7.1).

REFERENCES

Browning, A. J. 1963. Teaching and applying the gene-for-gene hypothesis for interactions in host:parasite systems. Proc. Iowa Acad. Sci. 70:120–125.

Flor, H. H. 1956. The complementary genic systems in flax and flax rust. Adv. Genet. 8:29–54.

Gold R. E.; Littlefield, L. J.; and Statler, G. D. 1979. Ultrastructure of pycnial and aecial stages of *Puccinia recondita*. Can. J. Bot. 57:74–86.

Lewis, D. H. 1976. Interchange of metabolites in biotrophic symbioses between angiosperms and fungi. In Perspectives in Experimental Biology. Vol. 2: Botany, ed. N. Sunderland, pp. 207–219. Elmsford, N.Y.: Pergamon.

Littlefield, L. J. 1972. Development of haustoria of *Melampsora lini*. Can. J. Bot. 50:1701–1703.

______. 1973. Histological evidence for diverse mechanisms of resistance to flax rust. Physiol. Plant Pathol. 3:241–247.

Littlefield, L. J., and Heath, M. C. 1979. Ultrastructure of Rust Fungi. New York, London: Academic Press.

Smith, D.; Muscatine, L.; and Lewis, D. H. 1969. Carbohydrate movement from autotrophs to heterotrophs in parasitic and mutualistic symbiosis. Biol. Rev. 44:17–90.

Stakman, E. C. 1915. Relation between *Puccinia graminis* and plants highly resistant to its attack. J. Agric. Res. 4:193–200.

Ziller, W. G. 1974. The Tree Rusts of Western Canada. Canadian Forestry Service Publication 1329.

FURTHER READING

Section 4.1 (Anatomy)

Arthur, J. C. 1929. The Plant Rusts (Uredinales). New York: John Wiley & Sons.

Boyce, J. S. 1961. Forest Pathology. New York, Toronto, London: McGraw-Hill.

Section 4.2 (Physiology)

Burrell, M. M., and Lewis, D. H. 1977. Amino acid movement from leaves of *Tussilago farfar* L. to the rust *Puccinia poarum* Neils. New Phytol. 79:327–333.

Goodman, R. N.; Kiraly, Z.; and Zaitlin, M. 1967. The Biochemistry and Physiology of Infectious Plant Diseases. Princeton, Toronto, London, Melbourne: D. Van Nostrand.
Heitefuss, R., and Williams, P. H., eds. 1976. Physiological Plant Pathology. Berlin, Heidelberg, New York: Springer-Verlag.
Chap. 1.2. Daly, J. M. Some aspects of host-pathogen interactions.
Chap. 5.2. Duniway, J. M. Water status and inbalance.
Chap. 5.3. Daly, J. M. The carbon balance of diseased plants: Changes in respiration, photosynthesis and translocation.
Keen, N. T., and Littlefield, L. J. 1979. The possible association of phytoalexins with resistance gene expression in flax to *Melampsora lini*. Physiol. Plant Pathol. 14:265–280.
Mendgen, K. 1979. Microautoradiographic studies on host-parasite interactions. II. The exchange of H^3-lysine between *Uromyces phaseoli* and *Phaseolus vulgaris*. Arch. Microbiol. 123:129–135.
Reisener, H. J., and Ziegler, E. 1970. Uber den Stoffwechsel des parasitischen Mycels und dessen Beziehungen zum Wirt bei *Puccinia graminis* auf Weizen. Angew. Bot. 44:343–346.
Whitney, P. J. 1977. Microbial Plant Pathology. New York: Pica Press.
Wood, R. K. S. 1967. Physiological Plant Pathology. Oxford, Edinburgh: Blackwell Scientific Publications.

Section 4.3 (Incompatibility)

Callow, J. A. 1977. Recognition, resistance, and the role of plant lectins in host-parasite relations. Adv. Bot. Res. 4:1–49.
Deverall, B. J. 1977. Defence mechanisms of plants. Cambridge, London, New York, Melbourne: Cambridge Univ. Press.
Heitefuss, R., and Williams, P. H., eds. 1976. Physiological Plant Pathology. Berlin, Heidelberg, New York: Springer-Verlag.
Chap. 1.2. Daly, J. M. Some aspects of host-pathogen interactions.
Chap. 8. Ellingboe, A. H. Genetics of host-parasite interactions.
Mendgen, K. 1978. Attachment of bean rust cell wall material to host and non-host plant tissue. Arch. Microbiol. 119:113–117.

5

Spore Biology

5.1 Release of Spores from Sori

Rust spores are set free from their hosts by a wide variety of mechanisms. Pycniospores are released from their supporting cells into a sugary, viscous liquid while still inside the subepidermal pycnia (see Chap. 2.1). This liquid matrix, containing thousands of pycniospores, forms droplets on the host surface as it oozes through the opening of the pycnium. Once on the host surface, the viscous liquid and spores are dispersed by insects, splashing water, and contact among leaves.

Aeciospores, produced in tightly packed chains (see Fig. 2.9), are set free as the alternating disjunctor cells between the spores disintegrate. In some rusts there are also large, dense granules on the aeciospore surface (Fig. 5.1). These are thought to aid in forcible ejection of spores from the chain. Once released from the aecium, aeciospores are wind dispersed. Moisture affects release of aeciospores differently in different rusts. In some *Gymnosporangium* spp. the long files of peridial cells (see Fig. 2.13) recurve downward when dry, facilitating release of the spores. In *Puccinia* spp. the peridium that encloses young aecia remains intact for several days at low humidity (see Plate 2.3), but when the humidity increases, the peridial cells rapidly absorb moisture and curve outward, causing rupture of the enclosing peridium (see Fig. 2.8). Aeciospores are then released into the air.

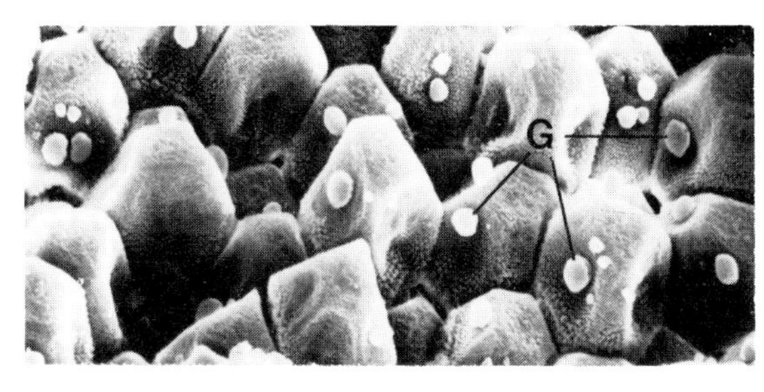

Fig. 5.1. Aeciospores of *Puccinia graminis* f. sp. *tritici* with minute verrucose ornaments and large granules (G) on their surface. ×1400. (From Littlefield and Heath 1979, by permission of Academic Press)

Urediniospores in most rusts are produced on pedicel cells within the uredinium, being exposed to the air once the host epidermis ruptures (see Fig. 2.15). As the urediniospores approach maturity, the junction between them and their subtending pedicels disintegrates, setting the spores free, whence they are wind dispersed.

Teliospores are variable regarding dispersal. The sessile teliospores of the Melampsoraceae and Coleosporaceae (see Figs. 2.19–2.21, 3.lb,c) remain attached to the host, where they eventually germinate to produce wind-borne basidiospores. In the Pucciniaceae, teliospores may germinate in place on the decomposing host plant tissue, or the pedicels on which they are borne (see Fig. 2.18b) break free near their base, remaining attached to the spore. The teliospores and their attached pedicels are wind dispersed together. Such teliospores may germinate wherever they land, if the environment is suitable, and there produce wind-borne basidiospores.

Basidiospores, because they are thin walled, rapidly dessicate compared to other wind-borne rust spores. Consequently, their range of effective dispersal is much less, e.g., about 300 m for *Cronartium ribicola,* except for some instances of 15–25 km dispersal. This contrasts to the effective range of thousands of kilometers for urediniospores.

5.2 Aerobiology

In this section will be considered the dispersal of infective rust spores by wind, which includes (1) local dispersal up to a few kilometers, (2) long-range annual dispersal over hundreds to thousands of kilometers, and (3) long-range dissemination of rust spores into areas previously free from the pathogen.

5.2.1 LOCAL DISPERSAL. Heavy infection of wheat with stem rust near barberry bushes furnished empirical evidence for spread of the pathogen from the bushes and provided the rationale for eradication efforts on barberry dating back to the seventeenth century. Convincing experimental evidence also came later from studies on white pine blister rust.

The pattern of dispersal of basidiospores of *Cronartium ribicola* and the resulting distribution of infected trees are markedly influenced by microclimatic effects caused by topography, water bodies, and vegetation as they affect convection currents of spore-laden air (Fig. 5.2). In the Great Lakes area of northern United States the currant alter-

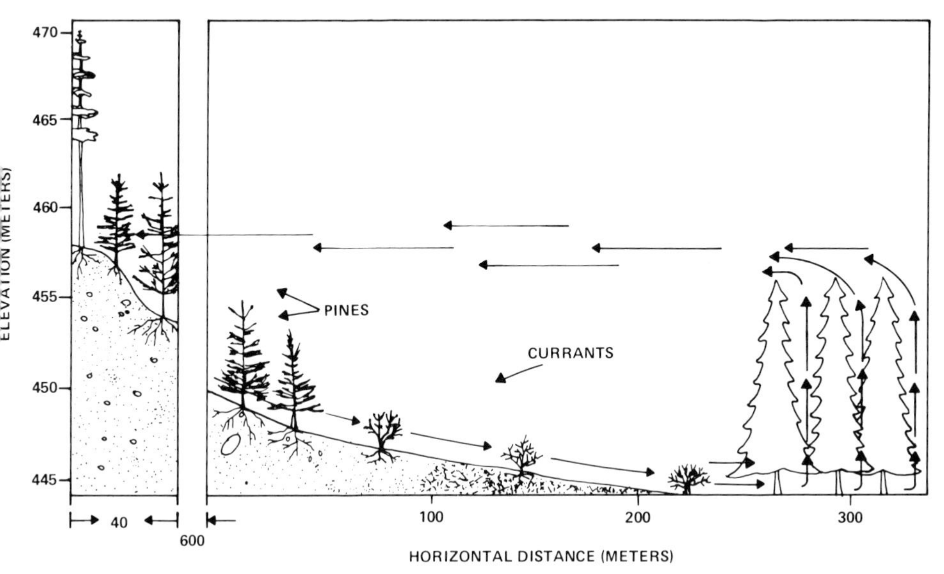

Fig. 5.2. Pattern of air flow that carries basidiospores of *Cronartium ribicola* originating on lower-elevation currant bushes to distant, upland white pines in northeastern Wisconsin, U.S. (Adapted from Van Arsdel 1965, by permission of the American Phytopathological Society)

nate host commonly inhabits low-lying areas around small lakes. The white pine host resides mostly on elevations above the many local depressions in which currants thrive. Basidiospores are released at night, particularly around 1 A.M., entering the stream of cool air that flows down from higher elevations into the valleys. These light-sensitive spores are small enough, about 10–12 μm diameter, not to settle out; they remain suspended in the air. The incoming cooler air rises over the relatively warmer water and flows back toward the land mass. This basidiospore-laden countercurrent flows horizontally until it eventually intersects the slope of the land and moves into the vegetation. Consequently, blister rust infection is most severe on pine trees several hundred meters uphill from the infected currants and least severe on those trees growing in low elevations close to the infected currants (Fig. 5.2).

Similar climatic events are also responsible for longer distance dispersal of *C. ribicola* basidiospores. On a peninsula separating Lakes Superior and Michigan in the United States blister rust incidence is highest some 15–25 km (10–17 mi) inland. There the basidiospore-laden countercurrent comes down to the land several kilometers from infected currants growing at lower elevations within 8–10 km (5–6 mi) of the lake shore (Fig. 5.3). Such effects help explain why eradication of currant bushes has proven ineffective as a sole method of controlling blister rust. Such dispersal of viable basidiospores over some 25 km is possible because it occurs at night. Otherwise, the fragile basidiospores of *C. ribicola* are killed within 5 hours when exposed to sunlight.

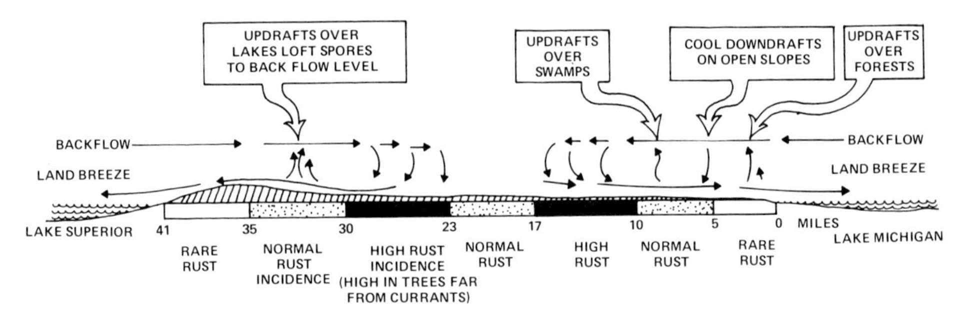

Fig. 5.3. Diagram showing the cross section of a peninsula between Lakes Superior and Michigan, U.S., the incidence of white pine blister rust, and the air flows controlling rust spread in that area. 1 mile = 1.6 km. (From Van Arsdel 1965, by permission of the American Phytopathological Society)

5.2.2 LONG-RANGE ANNUAL DISPERSAL. Long-range annual dispersal is typical of cereal rusts where millions of hectares of cereal grains are sown continuously over thousands of kilometers and where rapid asexual reproduction of the rust occurs in the uredinial stage. A most striking example of this is the "Puccinia path" in North America, extending some 4000 km from central Mexico to the Prairie Provinces of Canada. Northerly and southerly winds complement each other to maintain the process from year to year. Urediniospores originating on spring-sown grain in Canada and northern United States are blown southward in early autumn, there initiating infections on autumn-sown grain in the southern states. The rust overwinters primarily as uredinial stage mycelium in southern United States and Mexico. In May and early June, as urediniospore populations build and southerly winds increase, the urediniospores are blown northward in numbers adequate to initiate epidemics. As this wave of urediniospores moves northward it first initiates infection in autumn-sown wheats of the central states and eventually in spring-sown wheats of northern United States and Canada. The movement occurs in some years apparently by a succession of short jumps with intervening stops where inoculum multiples locally. This type of movement is facilitated by the continuous cultivation of susceptible genotypes over large geographic areas (see Chaps. 1.4, 7.5). In other years, if atmospheric and wind conditions are favorable, infection may appear almost simultaneously over large portions of the Puccinia path.

Similarly, in India stem rust survives the summer on cereal crops and volunteer plants in the cool, higher elevations of the north and Nepal. In autumn these spores are blown southward onto the principal wheat-growing plains, causing serious infections in autumn-sown crops.

In Europe *Puccinia graminis* f. sp. *tritici* originates annually from the south, although locally important foci originating from infected barberry are often important. Overwintering uredinial stage infections in Morocco and occasionally southern Spain and Portugal provide inoculum carried on southwesterly winds along the Atlantic coast, up to England, and possibly into Scandinavia. This is the so-called West European Tract of *P. graminis* dispersal. The East European Tract originates in the lower Danube plain in Bulgaria and Romania, probably from infected barberries. The rust proceeds westward along the Danube into Yugoslavia, Hungary, Austria, and Czechoslovakia, with the larger northern branch extending over the Ukraine, Poland, and into Scandinavia.

Yellow rust (*Puccinia striiformis*), in contrast, can develop severe, widespread epidemics throughout Europe without long-distance dispersal of inoculum by wind. Long-distance dispersal is important in distributing new races of this pathogen over wide areas of Europe, but once established, races of this cold-tolerant rust are not dependent upon annual inflow of large amounts of inoculum compared to *P. graminis.* They survive quite well on autumn-sown and volunteer wheat.

The numbers of spores involved in such mass air movements are astronomic. A wheat field moderately affected by stem rust can produce at least 25×10^6 urediniospores per m^2. A heavily rusted, 2 m tall barberry bush has been calculated to contain about 7×10^{10} aeciospores of *P. graminis.* Moreover, several crops of spores may be produced on a single bush during spring and early summer. Although less than 1% of these spores travel more than 100 m, their total numbers are phenomenal.

Once released into the air, turbulence can lift the spores hundreds to a few thousand meters, although spores become relatively scarce at altitudes above 3000 m. The vertical distribution of spores in air depends in part on the proximity of the region sampled to the source of the spores. In the case of locally produced spores, their vertical distribution approaches the ideal pattern of logarithmic decrease with height (Fig. 5.4, line A). In contrast, if spores are being transported from a distant source into an area not producing spores, e.g., in early summer in Canada, there are more spores at higher elevations than lower (Fig. 5.4, line B). In the latter case, the lower portion of the spore cloud was being depleted by such processes as rain washing and deposition by gravity.

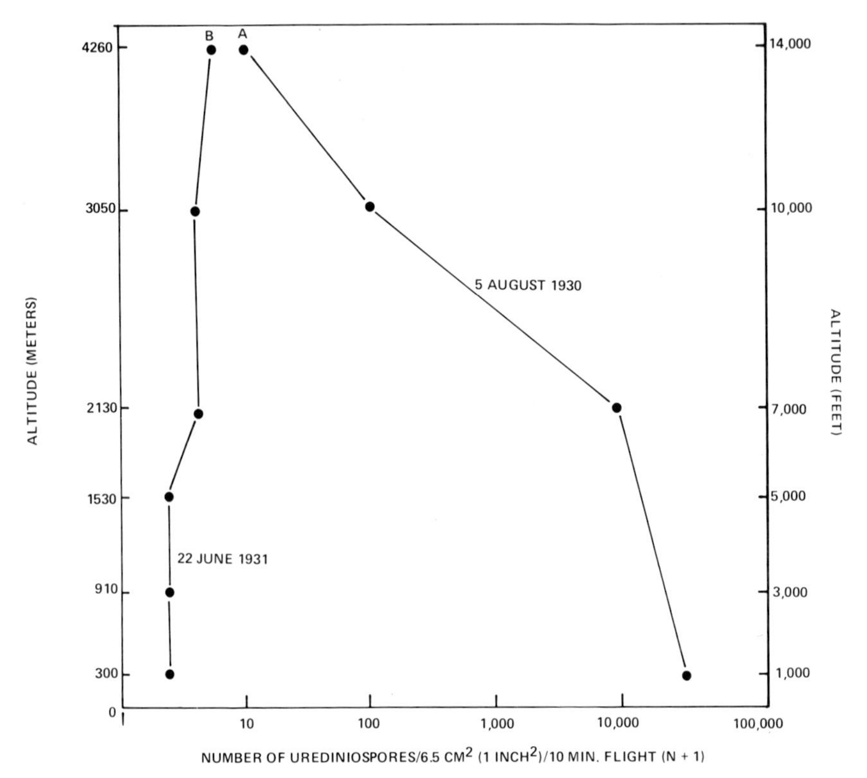

Fig. 5.4. Vertical distribution of *Puccinia graminis* f. sp. *tritici* urediniospores over southern Manitoba. (Line A) Late season data, collected 5 August 1930. (Line B) Early season data, collected 22 June 1931. (Adapted from Craigie 1945)

5.2.3 DISSEMINATION INTO NEW REGIONS. Natural barriers such as oceans and mountain ranges can drastically retard movement of rusts between geographic areas. Once the barrier has been crossed, however, either by wind dissemination (e.g., coffee rust) or inadvertent human introduction (e.g., tropical maize rust), winds may rapidly carry spores from the new focus over large areas. Figure 5.5 shows the spread of southern maize rust (*Puccinia polysora*) across Africa following its appearance in Sierra Leone in 1949. Simultaneously, the pathogen was spreading across Southeast Asia, occurring first in Malaya in 1948. Both introductions from tropical America are thought to have been by humans. Such rapid rates of spread are typical of wind-borne pathogens. Similarly, coffee rust (*Hemileia vastatrix*), once established in Brazil, has continued to spread across large areas on the prevailing northeasterly winds.

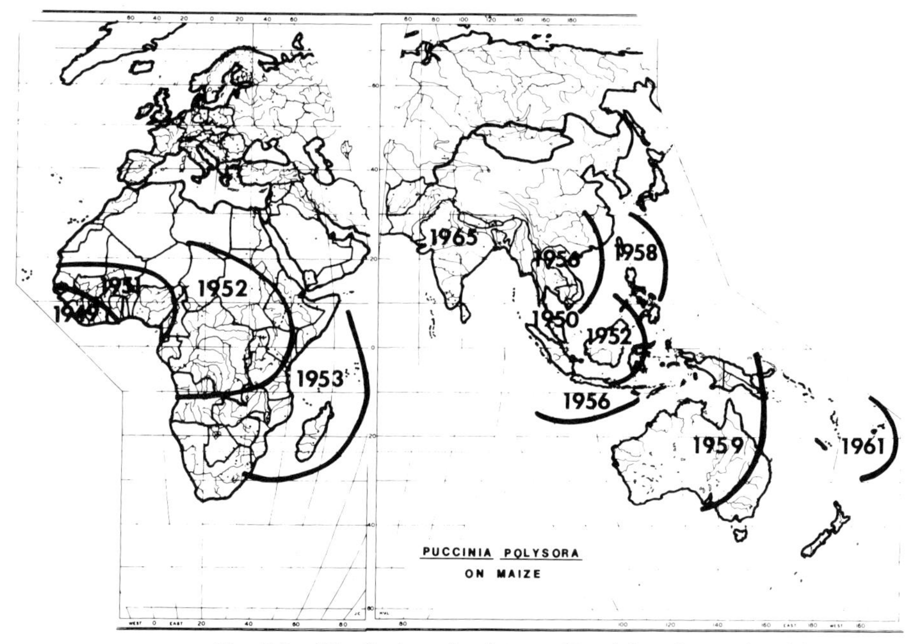

Fig. 5.5. Distribution of maize rust (*Puccinia polysora*) in Africa and Asia, 1949–1965. This pathogen, probably indigenous on maize in the Western Hemisphere, remained there until 1948. Once introduced into Malaya in 1948 it spread rapidly throughout Southeast Asia, Queensland, and the Pacific Islands. Similarly, it suddenly appeared in Sierra Leone in 1949 and spread across Africa within 5 years. (Adapted from Gregory 1973)

Melampsora medusae and *M. larici-populina,* long established pathogens of poplars and cottonwoods in many regions of the world, did not appear on the Australian continent until early 1972. Within 14 weeks, these newly arrived pathogens spread inland up to 400 km, extending from the north end of New South Wales southward to near Melbourne, Victoria (Fig. 5.6). The wet cyclonic weather during January and February coupled with the prevailing easterly winds was particularly favorable for the development and spread of these rusts. Within approximately 1 year of the discovery of these rusts in Australia, they were also found for the first time in New Zealand, more than 1600 km across the Tasman Sea. This spread appears also to have been by wind-borne urediniospores.

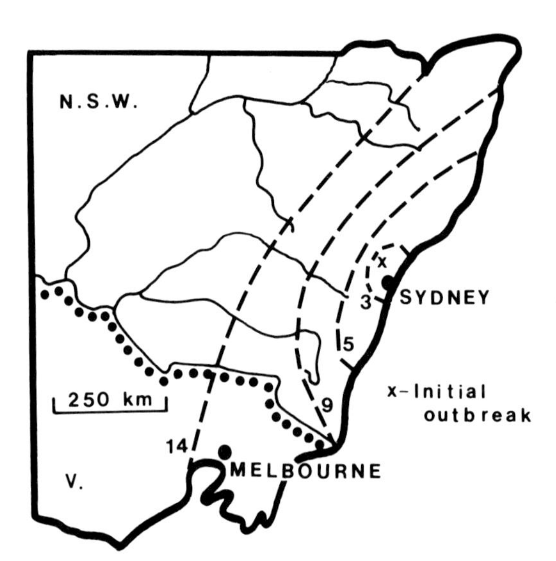

Fig. 5.6. Spread of *Melampsora medusae* into southeastern Australia in 1972 after 3, 5, 9, and 14 weeks. (N.S.W. = New South Wales; V. = Victoria) (Adapted from Walker et al. 1974, by permission of Verlag Paul Parey)

5.3 Dormancy and Germination

Urediniospores and teliospores of many rusts exhibit constitutive dormancy, i.e., a condition in which spores will not germinate until some internal barrier or inhibition has been removed, irrespective of how favorable environmental conditions may be.

Two chemical components of urediniospores have been identified as inhibitors of germination. They are methyl *cis* 3,4-dimethoxycinnamate (MCD) and methyl *cis* ferrulate (MF). Their effect is manifested in the germ pore region of the spore wall. Germination is normally accompanied by wall degradation within the germ pore, followed by growth of the germ tube through the resulting void, but the inhibitors prevent that essential first step of degradation of the cell wall. The inhibitors are water soluble and can be removed by soaking the spores in water.

There are also germination stimulators within the spore, e.g., B-ionone, which function to overcome the effects of the inhibitors. Consequently, successful germination depends in part on the critical balance between inhibitors and stimulators within the spore. The inhibitors function in minute concentrations in the range of some hormones. For example, the MCD inhibitor in peanut rust (*Puccinia arachidis*) is effective in the picogram range (pico = one-trillionth), with an ED_{50} of 8 picograms/ml. Most others are effective in the range of 1–10 nanograms/ml (nano = one-billionth).

Dormancy in teliospores can often be broken by various combinations of repeated freeze-thaw and wet-dry cycles or by heat shock, depending on the rust. The biochemical basis for these effects is not known.

Upon commencement of germination, before the emerging germ tube is visible, utilization of stored nutrient reserves also begins. In rust urediniospores the major nutrient reserves are lipids, composing some 20–25% of the dry weight. Carbohydrates, e.g., trehalose and mannitol, also function as nutrient reserves. The stored lipids are catabolized initially via B-oxidation to form acetyl-coenzyme A. The latter is metabolized via the Krebs cycle and the glyoxylate pathway along with other carbohydrates. A major function of the glyoxylate pathway in urediniospores may be to replenish 4-carbon acids used in biosynthesis of amino acids. The stored nutrients provide energy for growth and chemical building blocks for new cytoplasm, organelles, and cell walls until a parasitic relationship is established with the host.

Parasitism is thought to commence upon formation of the first haustorium, although absorption of nutrients from the host by the intercellular substomatal vesicle and infection hyphae before that time may occur.

REFERENCES

Craigie, J. H. 1945. Epidemiology of stem rust in western Canada. Sci. Agric. 25:285–401.

Gregory, P. H. 1973. The Microbiology of the Atmosphere, 2nd ed. Aylesbury, England: Leonard Hill.

Littlefield, L. J., and Heath, M. C. 1979. Ultrastructure of Rust Fungi. Chap. 2. New York, London: Academic Press.

Van Arsdel, E. P. 1965. Micrometerology and plant disease epidemiology. Phytopathology 55:945–950.

Walker, J.; Hartigan, D.; and Bertus, A. L. 1974. Poplar rusts in Australia with comments on potential conifer rusts. Eur. J. For. Pathol. 4:100–118.

FURTHER READING

Section 5.2 (Aerobiology)

Ingold, C. T. 1971. Fungal Spores, Their Liberation and Dispersal. Oxford: Clarendon Press.

Stakman, E. C., and Harrar, J. G. 1957. Principles of Plant Pathology. New York: Roland Press.

Zadoks, J. C. 1965. Epidemiology of wheat rusts in Europe. FAO Plant Prot. Bull. 13:97–108.

Section 5.3 (Dormancy and Germination)

Weber, D. J., and Hess, W. M., eds. 1974. The Fungal Spore, Form and Function. New York, London, Sydney, Toronto: John Wiley & Sons.

Chap. 2. Macko, V.; Staples, R. C.; Yaniv, Z.; and Granados, R. R. Self-inhibitors of fungal spore germination.

Chap. 4. Gottlieb, D. Carbohydrate metabolism and spore germination.

Chap. 5. Reisener, H. J. Lipid metabolism of fungal spores during germination.

6

Axenic Culture of Rusts

Rusts have long been defined as obligate parasites, unable to grow separately from their host plant. This description remained valid until the 1950s and 1960s when several species were cultured in the laboratory on chemically defined media. However, rusts have no known saprophytic ability in nature. Consequently, "obligate parasites in nature" or "ecologically obligate parasites" are still accurate designations of rust fungi.

The term "axenic culture" (a + Gr. *xenos* = strange; hence, free from other living organisms) is used when referring to cultures of rusts isolated from the host, growing in the mycelial condition on synthetic medium, in the complete absence of other organisms. It is not to be confused with "pure culture" of rust, sometimes used to connote a single species or physiologic race of rust growing upon its host.

Axenic cultures of rusts have been produced by various means. The first fully documented study was by Hotson and Cutter in 1951, followed in detail by Cutter in 1959, working with *Gymnosporangium juniperi-virginianae* in the United States. After years of exhaustive study, Cutter obtained seven rust isolates in culture from some 14,000 telial gall slices. Upon reinoculation of the mono- and dikaryotic mycelia into the juniper and hawthorn hosts, he obtained infection and subsequently aecial and telial sori on the respective hosts. Unfortunately Cutter did not live to see his work duplicated, and his results were often questioned. In the 1970s, infective axenic cultures were obtained with similar methods on aecial galls from *Cronartium fusiforme* and *C. ribicola*.

In 1966, Williams et al. in Australia reported the first axenic culture of *Puccinia graminis* f. sp. *tritici* by direct "seeding" of aseptic urediniospores onto an agar medium. This procedure gave rise to mycelial cultures on a simple mineral medium containing yeast extract and sucrose in agar. This work was repeated by numerous North American and European workers with additional races of stem rust as well as flax, sunflower, carnation, bean, and wheat leaf rusts. At last the validity of Hotson and Cutter's work was recognized. A good explanation is still needed for why so many years of fruitless effort were expended in attempts to culture the rusts, and then success was obtained on such a simple nutrient medium.

A third method of obtaining axenic rust cultures is to digest infected leaves with cell wall–dissolving enzymes to release the mycelium, which can then be transferred to a culture medium. This has been used successfully with flax rust.

Nutritionally, the demands of axenic rust cultures are unpredictable and varied. Often cultures derived from urediniospores are more fastidious than are subcultured isolates. For example, in one isolate of flax rust, fructose and arabitol are utilized by subcultured isolates but not by those grown directly from urediniospores. Numerous carbohydrates, e.g., sucrose, glucose, fructose, mannose, trehalose, raffinose, mannitol, ribitol, and sorbitol, among others, can serve as carbon sources for various axenically cultured rusts. There appears to be little specificity in carbohydrate requirement.

Organic nitrogen can be provided by casamino acids and peptone. Although many amino acids can be utilized, some are better sources of nitrogen than others. In flax and sunflower rusts the sulfur-containing amino acids cysteine or cystine must be present for many of the other amino acids to be utilized. In fact, sulfur amino acids are required by all rust fungi grown on a chemically defined medium. This is thought to result from a block in the metabolism of inorganic sulfate, which prevents reduction of the latter to thiosulfate or sulfide. This requirement for sulfide sulfur or sulfur-containing amino acids is interesting in that sulfate reduction occurs in bacteria, blue green algae, green algae, other fungi, and higher plants. Neither nitrate nor nitrite nitrogen can be utilized by cultures of *Puccinia graminis.*

Morphologically, rusts produce aeciospores, urediniospores, or teliospores depending upon what stage of the life cycle is being cultured. Pycniospores have not been produced in culture. Axenically produced teliospores of *Gymnosporangium juniperi-virginianae* can germinate to form basidia and basidiospores, as they do in nature. Axenically produced aeciospores and urediniospores of *Melampsora lini* when placed onto flax leaves can germinate, form infection structures, and establish a normal parasitic relationship with the host. However, with urediniospores of *P. graminis* or with mycelia of various rusts grown axenically, the host epidermis must be lifted up and the fungus then placed directly onto the exposed mesophyll for infection to occur. The reason(s) for such a requirement by axenically produced urediniospores is uncertain.

Initially, great expectations were expressed for the scientific potential of axenic rust culture research. It was hoped that axenic cultures could be used to circumvent the lengthy, laborious procedures required for studying genetics of the fungus grown on the host plant. Also, axenic cultures should be an ideal system for studying metabolism and nutrition of rusts, separate from their hosts. However, their exceedingly slow growth rate and unpredictable nutritional requirements have provided serious obstacles to the widespread use of axenic cultures in rust research.

REFERENCES

Cutter, V. M., Jr. 1959. Studies on the isolation and growth of plant rusts in host tissue cultures and upon synthethic media. I. *Gymnosporangium.* Mycologia 51:248-295.

Hotson, H. H., and Cutter, V. M., Jr. 1951. The isolation and culture of *Gymnosporangium juniperi-virginianae* Schw. Proc. Natl. Acad. Sci. 37:400-403.

Williams, P. G.; Scott, K. J.; and Kuhl, J. L. 1966. Vegetative growth of *Puccinia graminis* f. sp. *tritici,* in vitro. Phytopathology 56:1418-1419.

FURTHER READING

Coffey, M. D. 1975. Obligate parasites of higher plants, particularly rust fungi. Symp. Soc. Exp. Biol. 29:297-323.

Scott, K. J. 1976. Growth of biotrophic parasites in axenic culture. In Physiological Plant Pathology, ed. R. Heitefuss and P. H. Williams, Ch. 7.1. Berlin, Heidelberg, New York: Springer-Verlag.

Scott, K. J., and MacLean, D. J. 1969. Culturing of rust fungi. Annu. Rev. Phytopathol. 7:123-146.

7

Control of Rust Diseases

Prevention is obviously the best control measure for any disease, rusts included. Where this is not possible, other means must be employed. Any single method of disease control has its limitations; consequently, an integrated approach utilizing several means of control is preferred. In this chapter are discussed some of the more important means of controlling rust diseases.

7.1 Eradication

Eradication refers to the elimination of a pathogen from an area into which it has been introduced or occurs naturally. It is not to be confused with sanitation, a cultural practice (see following section) aimed at removing diseased host material (e.g., pruning, burning leaves) to reduce the amount of inoculum present. Other procedures such as quarantines may be directed against the introduction of pathogens into areas where they are not currently found.

Since rusts are obligate parasites in nature, eradication of their hosts should theoretically eliminate the pathogen. Numerous problems, however (e.g., widespread wind dispersal of spores, survival of cut stumps, and wide host ranges), limit successful eradication programs in many cases.

Eradication may be aimed at either the primary host of autoecious rusts (where an alternate host is yet unknown) or at the alternate host(s) of heteroecious rusts. Examples of the former are few, although two very effective campaigns can be cited. Coffee rust, widespread over eastern Africa and southern Asia by 1900, was inadvertently introduced into Puerto Rico in 1903. That island location, far removed from natural sources of airborne inoculum, was well suited to an eradication effort. By quick and thorough removal of infected trees, the pathogen was destroyed in Puerto Rico within 1 year. The Western Hemisphere then remained free of coffee rust until 1970. Similarly, a recent success can be cited in Papua in 1965. There, a prompt, intensive campaign eradicated coffee rust within 1 year. Success there was aided by the isolated locations of initial infections and their separation by miles of rain forests and mountains from the major coffee-growing regions. In all cases the costly eradication and replanting programs were more than compensated. The cost of the 1965 campaign in Papua was $70,000 (Australian), and the value of exported coffee 1965–1973 was $143 million (Australian). Much of that export income would have been lost had the rust not been eradicated.

Eradication of alternate hosts has been a more widespread practice. In no case has a rust fungus or one of its alternate hosts been totally eradicated, but frequently the economic impact of the disease has been reduced significantly. The oldest example of eradication of an alternate host is that of barberry, initiated in France in 1660 (see Chap. 1). In North America an extensive barberry eradication program, eventually covering about 1 million square miles (2.6 million km^2), was initiated during World War I. That program was brought about partly by the calamitous stem rust epidemic of 1916 and the increased demands for wheat because of the war.

The area included 19 states in the United States and 3 adjacent provinces in Canada. The cost for eradicating 500 million barberry bushes was great, but in the United States an annual savings of over $30 million has been estimated for that program. Barberries (except rust-resistant horticultural varieties) are still outlawed in those states. Benefits of that eradication effort, although now terminated because of its thoroughness, are still being derived. The benefits come both from the elimination of early spring inoculum—hence delayed disease initiation—and from the prevention of new physiologic races that can be formed by hybridization of the rust fungus on the barberry host.

Barberry eradication in Denmark, initiated in 1903, has been successful in preventing stem rust epidemics that originate there from local aeciospore inoculum. Barberry eradication programs in Bavaria, started in the 1950s, have given good results. In contrast, large numbers of barberry render some Swiss alpine valleys practically useless for growing wheat. Similarly, a recent rust epidemic in Ireland was traced to the many barberry hedges and bushes in the southern part of the country.

Although barberry occurs in Great Britain, it does not help to perpetuate wheat stem rust. Inoculum for stem rust there comes annually from the Iberian peninsula and northern Africa, being wind borne (see Chap. 5.2.2). Barberry rarely occurs in Australia, with inoculum there coming from volunteer wheat or other grass hosts. Thus plowing down of those alternative hosts (not "alternate" hosts) is often used to control stem rust.

Another large-scale eradication effort was aimed at currant (*Ribes* spp.), an alternate host of *Cronartium ribicola,* the white pine blister rust pathogen. That program commenced in 1930 in the United States. Basidiospores of this fungus (produced upon germination of sessile teliospores on currant) are short-lived compared to the durable aeciospores of *Puccinia graminis* on barberry. Consequently, eradication of currant need not be as extensive as that of barberry. Removal of currants within a 350 m distance of white pine stands is recommended. Originally, this removal was by hand, but later they were removed by herbicides. Although white pine blister rust was lessened somewhat by this effort, the U.S. Forest Service has now terminated this program because of economic considerations and the inability to reduce currant populations below the critical level necessary to prevent pine infection.

Eradication of the juniper alternate host of *Gymnosporangium fuscum* has been successful in retarding spread of pear trellis rust in British Columbia. Although introduced from Europe in the early 1930s, it was restricted to the lower tip of Vancouver Island until the early 1960s, primarily because of extensive eradication of teliospore-bearing junipers. It has since become distributed over much of coastal British Columbia, but during 1975–1978 more than 1000 infected junipers were removed, with an accompanying 93% reduction in pear infections. Simultaneously, a program now exists to guarantee release to homeowners only junipers certified to be rust free.

7.2 Cultural Practices

Rotation can be beneficial in controlling autoecious rusts, e.g., bean and flax rusts. Overwintering teliospores of most temperate zone rusts will not survive more than one season in contact with the soil. If rotation is not practicable, then burning infected debris or deep plowing will reduce the amount of inoculum available for the succeeding crop. Removal of infested, inoculum-bearing debris is a standard sanitation procedure for rust control in gardens and glasshouses, e.g., rusts of asparagus, hollyhock, geranium, iris, and snapdragon. Similarly, deep autumn plowing of peppermint is recommended in control of *Puccinia menthae.* This practice buries teliospore inoculum that would otherwise produce basidiospores to initiate infection of newly formed shoots from stolons the following spring.

A more effective control of this rust, however, is flame sanitation, in which fields are flamed by a propane torch in the spring. This kills young lesions initiated by basidiospore infection and precludes formation of subsequent aeciospore and urediniospore inoculum. Pruning out inoculum-producing galls of the autoecious western gall rust (*Endocronartium harknessii*) is effective in isolated plantings, e.g., nurseries and farmstead windbreaks in the U.S. Great Plains.

Reducing the relative humidity will also help limit rust development. A common, inexpensive control of geranium and carnation rusts in glasshouses is to water plants by surface irrigation, rather than by sprinkling, in combination with adequate ventilation. In addition to reducing relative humidity, this also prevents dispersal of urediniospores by the sprinkling water.

7.3 Chemical Treatments

Fungicides, depending on their chemical structure and mode of action, may be applied to the surface of the plant or to the soil. They may remain on the plant surface or be absorbed and thence translocated systemically throughout the plant. Fungicides that function on the plant surface to inactivate the fungus prior to its entry are termed protectants. Those that are applied subsequent to the establishment of infection and then kill the fungus are termed eradicants. Compounds that remain on the surface of the plant most commonly function as protectants. Systemic fungicides, depending on the specific compound, may be protectants and/or eradicants but commonly are the latter. Some compounds actually kill the fungus; others only arrest further growth. These are termed fungicidal and fungistatic in their action respectively.

Over the past five decades, much research has been devoted to chemical control of cereal rusts. Inorganic sulfur was one of the first protectant fungicides to be used. Research in the 1920s and 1930s demonstrated the economic justification of repeated applications of sulfur in years favorable for heavy rust infection. However, fungicidal control of cereal rusts was not utilized extensively for economic reasons. Many compounds were tested and found to be effective in glasshouses and small test plots, but they were unsuited for field use. They were either too phytotoxic, too costly, or insufficiently active. Such materials included borax, picric acid, lithium salts, lime sulfur, maleic hydrazide, and various sulfonamides. In contrast, the inorganic protectant copper oxide provides effective, economic control of coffee rust in plantations.

More recently, organic fungicides, both the sulfur-containing dithiocarbamate surface protectants and various systemic protectants or eradicants, have been used economically for cereal rust control. The systemic compounds include oxycarboxin, effective against essentially all rusts, and RH-124 (4-n-butyl-1,2,4-triazole), which curiously is effective only against wheat leaf rust (brown rust in the United Kingdom). The latter compound is fungistatic, not fungicidal. RH-124, effective as a foliar eradicant, is also active when applied as a seed treatment and will protect against rust development for several weeks after planting.

Economic variables that dictate the value of such applications include cultivar susceptibility, potential yield, probable favorableness for disease in the environment, price of the wheat, and rust severity. All these values must be known or closely estimated before making the decision to spray. Considering these variables, the expected return for two spray applications of protectant fungicide for wheat leaf rust control in North Dakota in 1979 are given in Table 7.1. Some of that benefit, however, would come from controlling the Septoria and other foliar pathogens simultaneously with the target fungus.

TABLE 7.1. Net profits from controlling foliar diseases of wheat with fungicides

Expected Loss			
Percent	Bushels per acre	Kg per hectare	Net Return per Acre* (dollars, U.S.)
0	0	0	(−8.50)
10	4	269	5.50
20	8	538	19.50
30	12	807	33.50

Source: North Dakota Agricultural Experiment Station, 1979 figures.

* Net return is expected yield loss (bu/A) times $3.50 per bushel minus the cost of two applications of the fungicide ($8.50/A), with estimated potential yield of 40 bu/A.

The systemic compounds oxycarboxin, triforine, and benodanil; the surface protectants zineb, maneb, and ferbam; as well as inorganic sulfur are variously recommended for numerous glasshouse, garden, and orchard crop rusts.

Chemical control of conifer rusts on an economic scale is more difficult, although oxycarboxin has been used successfully on needle rust of fir Christmas trees in northwestern United States. Limited control of pine fusiform rust is possible with several compounds, although their use is limited mostly to eradicating galls in seed orchards and ornamental plantings.

The somewhat systemic antibiotic cyclohexamide (Actidione) was released in the 1940s as a rust eradicant. This material was applied extensively for the control of white pine blister rust in the United States during the late 1950s and early 1960s. The effort proved in vain, however; after initial success the favorable results were not duplicated. Eventually, Actidione was shown to be effective only if applied directly to scarified cankers; also, dosage levels translocated to other parts of the tree were insufficient to control the disease. No formulation was found that killed the fungus without also killing the tree. Ironically, the widespread application of Actidione may have done more damage than good. Evidence exists that *Tuberculina maxima,* a fungal parasite of *Cronartium ribicola* (see next section), was inhibited more than the latter by Actidione, and the resulting degree of biological control that *T. maxima* provided in nature was diminished.

A type of chemical control, strictly coincidental and unplanned, is that provided by various industrial air pollutants. For example, SO_2 has been found to reduce the severity of wheat stem rust and numerous conifer rusts in industrialized areas of the United States, Canada, and Sweden. Both rusts and wood-decay fungi are uncommonly sensitive to SO_2 compared to other fungi. Ozone reduces the severity of oat crown rust and wheat stem rust, as does fluoride for bean rust. The latter results have been demonstrated only under experimental conditions, however. Although such responses may appear to be positive effects of air pollution, it must be remembered that rust fungi comprise only a small proportion of the biota. Also, little is known of the overall effects of such alterations of natural ecosystems and the overall consequences.

7.4 Microbiologic Agents

Microbes deleterious to the growth and reproduction of rust fungi may act by direct parasitism of the rust, indirect injurious effects on the host, or production of antibiotics. Other modes of action may also be involved, e.g., competition for nutrients and/or space. The associated organisms include fungi, bacteria, and viruslike particles. Fungi that parasitize other fungi are known as mycoparasites. If the mycoparasite happens to parasitize a parasitic fungus, it is then often referred to as a hyperparasite. As yet, there is only the potential for microbiologic control of rusts. No examples of successful commercial application can yet be cited.

More than 30 genera of fungi have been found inhabiting sori on rust-infected plants, but how many of these are genuinely parasitic on the rust is uncertain. One of the most commonly encountered rust parasites, *Darluca filum,* occurs in uredinia. The mycelium of this fungus grows among and sends branches into the urediniospores, destroying cytoplasm and eventually killing the infected spores. Similar colonization of uredinia and urediniospores occurs with *Alternaria* spp. and *Cladosporium* spp., both common leaf surface saprophytes. Experimental studies show that when applied to poplar leaves prior to inoculation with *Melampsora larici-populina,* these two genera significantly reduce urediniospore germination and subsequent development of the rust infection. *C. gallicola* similarly invades aecioid teliospores of *Endocronartium harknessii* (North American western gall rust on pines). It digests the cytoplasm and eventually forms conidia and conidiophores on the parasitized spores (Fig. 7.1). *Verticillium lecanii,* another asexual fungus, is often parasitic on urediniospores of *Uromyces dianthi* (Fig. 7.2) and *Puccinia striiformis,* although extensive colonization of the latter does not develop at relative humidity less than 90–95%. This hyper-

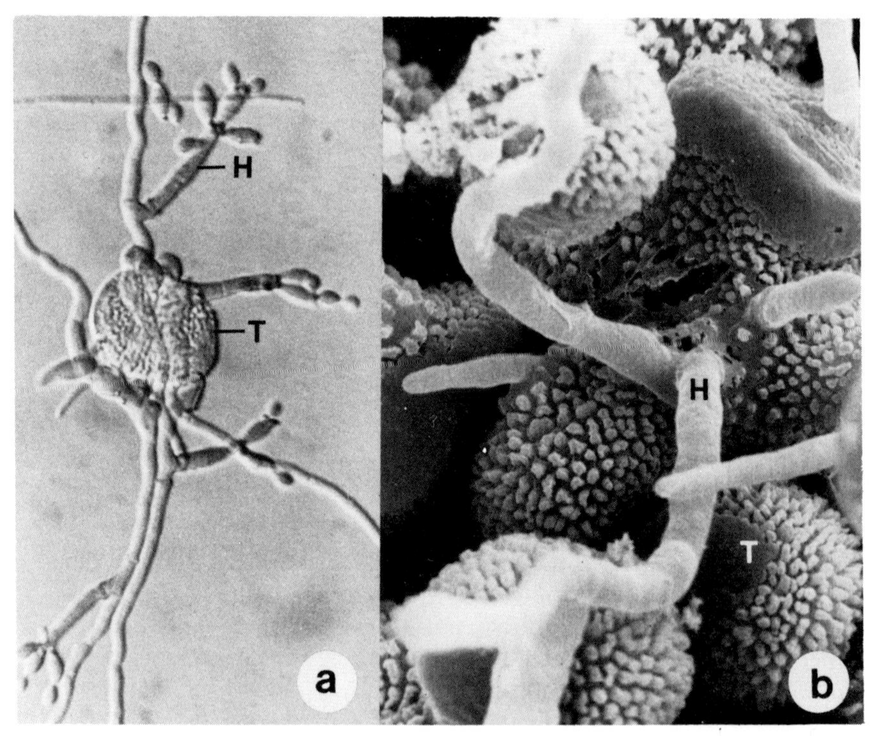

Fig. 7.1. Parasitism of aecioid teliospores (T) of *Endocronartium harknessii* by hyphae of *Cladosporium gallicola.* (a) Light microscopic view shows development of asexual reproductive spores by the hyperparasite (H). ×1200. (b) Scanning electron micrograph shows emergence of the hyperparasite's hyphae (H) from the rust teliospores (T). ×2500. (From Tsuneda and Hiratsuka 1979, by permission of the Canadian Phytopathological Society)

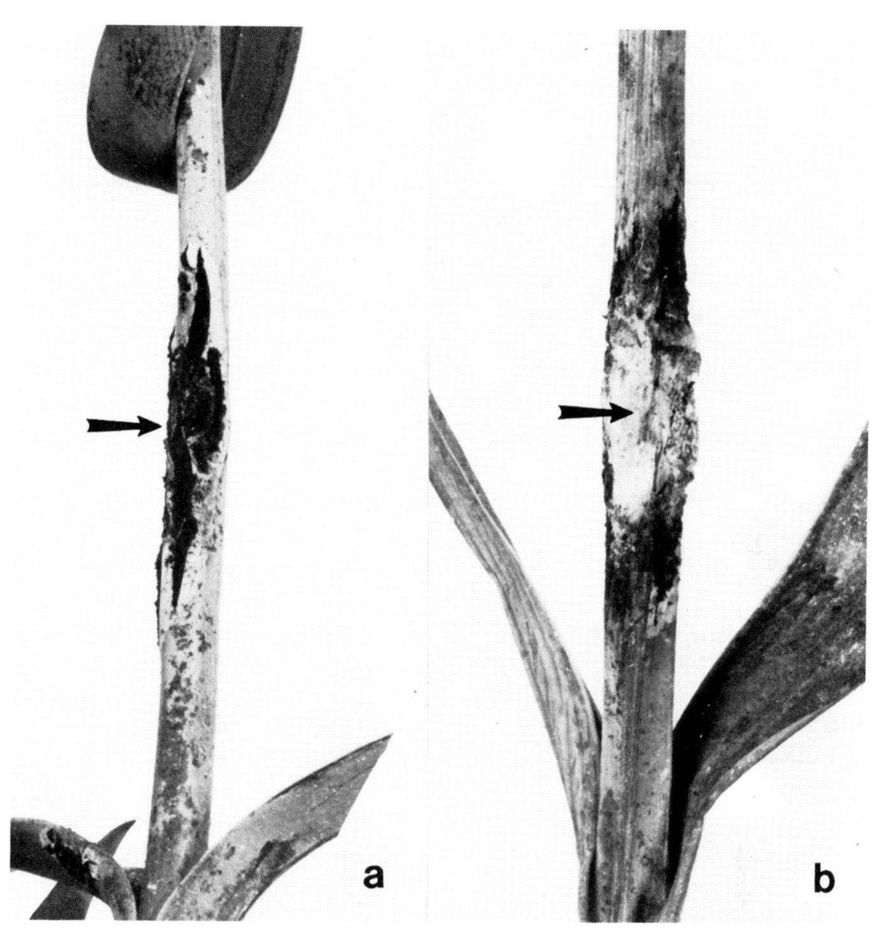

Fig. 7.2. Inhibition of the development of uredinia of *Uromyces dianthi* on stems and leaves of carnation when the rust fungus is infected by the hyperparasitic fungus *Verticillium lecanii.* (a) Normal development of dark uredinial sori on stem and leaves (arrow). (b) *V. lecanii* (light region, arrow) growing on coalesced uredinia of the pathogen. (From Spencer 1980, by permission of Cambridge University Press)

parasite does not invade the *P. striiformis* urediniospores. Rather, it destroys the cytoplasm by lytic action and eventually digests the urediniospore wall, leaving only the surface spines intact.

An indirect effect of the antagonistic fungus on the rust rather than direct parasitism per se has been described in several instances. *Typhula idahoensis* and *Fusarium nivale,* which cause snow mold, can destroy overwintering urediniospore inoculum of *Puccinia striiformis* in the Pacific Northwest area of North America. Snow mold indiscriminately destroys wheat and other grass leaves underneath the snow cover, which results in destruction of resident rust spores as well. In the spring, those plants renew growth from underground crown tissue and are rust free.

In that same region of North America, aecial cankers of white pine blister rust (*Cronartium ribicola*) are rendered inactive by *Tuberculina maxima.* Uninfected tissues of western white pine are not susceptible to *T. maxima.* The hypertrophied aecial cankers are susceptible; this has been suggested as resulting from the absence of pectin in cell walls of such cankers. Once the antagonistic *T. maxima* invades the canker tissue the host cells are killed, and the rust dies. The canker is thus rendered ineffective. The rust mycelium and spores are not invaded. Such antagonism is important in limiting expansion of the canker, thus preventing girdling and death of branches and extension of the canker into the main stem. Also, the number of aecia per canker is greatly decreased, limiting inoculum available for initiating other infections. Recent studies have shown about two-thirds of the lethal-type cankers present in natural stands were inactive and that *T. maxima* was fruiting on about only one-fourth of those. Aeciospore production, i.e., the potential inoculum, was reduced by about 96% in the population of lethal-type cankers studied. This may well be one example of a highly effective microbiologic control of rust in nature.

Three species of *Bacillus* that occur commonly as saprophytes on needles of Douglas fir when applied as a mixture onto needles of glasshouse-grown seedlings inhibit germination of basidiospore inoculum of *Melampsora medusae* and control that rust. The mechanism of this antagonism is unknown.

Viruses—or viruslike particles (VLPs), to be more correct, since unequivocal proof of their virus nature is lacking—occur widely in fungi. They have been observed in various stages of the following rust fungi: *Uromyces phaseoli, Puccinia helianthi, P. graminis* f. sp. *tritici, P. malvacearum, P. striiformis, P. suaveolens, Hemileia vastatrix,* and *Coleosporium asterum.* The VLPs observed consist of spherical and rod-shaped forms of various dimensions. To date, the spherical VLPs observed in the rusts appear to correspond morphologically to the double-stranded RNA viruses that occur throughout the fungi. However, the rod-shaped VLPs observed do not resemble morphologically other such viruses reported in fungi. Exceedingly little is known about viruses in the rusts.

Except for the demonstrated action of *Tuberculina maxima* against

Cronartium ribicola in the field and perhaps *Darluca filum* against various *Puccinia* spp., microbiologic control of rust diseases is far from being developed as a viable means of rust control. However, as fossil fuel supplies become further diminished, as regulations on pesticide usage become more restrictive, and as public concern over environmental pollution increases, there may develop a greater interest in the use of microbiologic controls against rust diseases.

7.5 Plant Breeding

Long before Mendel, astute agriculturalists and early plant breeders were empirically crossing plants and making selections to obtain superior lines. As early as 1796 some hybrids of wheat developed by Knight in England were free from "blight" (probably caused by *Puccinia striiformis*). In the 1880s, Farrer produced numerous rust-resistant wheat hybrids in Australia, and by 1894 Hitchcock and Carleton in the United States recognized that the breeding of "rust-proof" wheats provided the best means of rust control.

The scientific foundation for subsequent disease resistance breeding research, however, was not laid until 1905 by Biffen, working in Cambridge, England. Biffen demonstrated that resistance to yellow rust was inherited in a Mendelian fashion. In segregating the progeny of a cross between a very susceptible and a resistant variety, disease resistance was shown to be controlled by a single recessive gene. Later, resistance was shown to be conditioned more commonly by dominant genes. Soon extensive breeding programs were initiated in many countries, and numerous rust-resistant cultivars of many crop plants were produced. To date, some 35 single genes for wheat stem rust resistance and some 30 for flax rust resistance have been identified.

An essential first step in any disease resistance breeding program is to obtain sources of resistance. This can be done in a variety of ways. Selection of resistant individuals from a population of diseased plants within cultivars has long been used. For example, two of the earliest stem rust–resistant bread wheats in North America, Kota and Kanred, were derived from selections of imported, diverse Russian wheats. Although later supplanted by other cultivars, Kota and Kanred contributed resistance genes that were utilized in breeding programs for many years.

Other sources of resistance that have proved useful exist in closely related wild or domesticated species. Pioneering research in this field was done by Vavilov, the great Russian geneticist, who after extensive botanical exploration identified natural centers of origin of many crop plants. Because wild species in those centers of ancient origin have been exposed to selection pressures for thousands of years, the likelihood of their containing diverse genes for disease resistance is high. Once the sources of resistance are identified, then hybrids between them and related agronomic plants are made to transfer the resistance gene(s) from the wild plant species to domesticated ones.

Also, intergeneric crosses have been used to transfer rust resistance genes to wheat from the genera *Secale* and *Agropyron*. Such procedures are exceedingly difficult at best and may require many years of research. Major problems include sterility in such attempted crosses, or when they are fertile the rust resistance genes are often linked to undesirable traits.

Problems can arise even when rust resistance genes are transferred from other cultivars of the same species. For example, cultivars of oats resistant to all known races of *Puccinia coronata* (crown rust) were released in the United States in 1942. Their resistance was derived in part from the cultivar Victoria, which came from Uruguay. By 1947 these cultivars occupied 92% of the Iowa oat acreage and quickly brought crown rust under control. Unfortunately, unknown to its breeders, the gene Pc-2, which conditioned crown rust resistance in these new cultivars, also conditioned susceptibility to *Helminthosporium victoriae,* a previously unknown minor pathogen of neighboring wild grasses. This fungus readily attacked the new "Victoria cultivars," and by 1946 and 1947 the "Victoria blight" ran rampant over vast areas of north central United States.

Other sources of disease resistance genes are mutations, either natural or induced. Chemical- or radiation-induced resistance to rusts has been utilized in several new cultivars of cereals released in Argentina, India, the United States, and Europe.

As discovered by Biffen with wheat yellow rust, most plants possess genes specific for resistance to various rusts. Moreover, this specificity extends to the level of physiologic races. For each gene in the fungus that conditions pathogenicity, there is a corresponding gene in the host that conditions the response of the plant to that corresponding gene for pathogenicity in the fungus (see Chap. 4.3.1).

Such resistance has been termed "race-specific" or "vertical"; it has been the primary focus of breeding for rust resistance since early in this century. The success obtained initially was spectacular. Equally spectacular was the demise of previously "resistant" cultivars when a new gene(s) for pathogenicity arose in the rust population for which there was not a corresponding gene(s) for resistance in the host. Consequently, resistant cultivars were successively rendered useless, and constant efforts have been required to stay ahead of the genetic plasticity of the rust fungi.

New approaches attempting to simulate resistance as it occurs in nature are being devised. One is the use of "multiline varieties," which were developed in Mexico, Colombia, India, and Iowa by international teams of breeders and pathologists. A multiline variety is a mechanical mixture of some 8–16 phenotypically similar lines that differ genotypically for rust resistance. Ideally, such a heterogeneous plant population will prevent the buildup of an epidemic even when there are shifts in the genotype of the rust population. The first two multiline varieties of wheat were released and grown successfully in Colombia. In North America 13 multiline oat varieties have been released to buffer against the damaging crown rust fungus. Some 400,000 hectares of such varieties of oats have been planted annually in Iowa and adjoining states so far without report of crown rust damage. Such varieties were recently released in India and Pakistan for control of yellow rust and leaf rust of wheat.

Another means of "managing" rusts, yet employing race-specific resistance, is to restrict the planting of race-specific cultivars to limited geographic areas. For example, in the "Puccinia path" of North America (see Chap. 5.2), progress is being made toward planting oat cultivars having different crown rust resistance genes in the different latitudes in which oats are grown. Thus as individual urediniospores blow northward in the spring, they are confronted by different oats that are each resistant to them. Large populations of a pathogenic race would thus be thwarted in developing and extending a crown rust epidemic throughout the entire north-to-south range of oat culture in North America, as happened in the decades of 1940–1960. Major problems encountered with this system are the political and economic decisions that must be made and imposed on independent breeders and growers.

A very different approach, one that gets away from simply managing race-specific genes, is the use of "race-nonspecific" resistance. Such resistance is typical in wild plants in areas of their origin, where this type of so-called "horizontal" or "generalized" resistance evolved over thousands of years. Race-nonspecific resistance has been identified in numerous agronomic crops and is expressed as partial susceptibility, with little accompanying yield loss. It is not as spectacular nor as readily identifiable as race-specific resistance, but theoretically its permanence is more enduring. In fact, the British refer to it as "durable resistance." The rationale for its use is that growers are better off economically to accept a low level of rust infection and yield loss year in and year out than be subjected to the successive "feast or famine" that accompanies use in pure stands of highly race-specific cultivars and their successive attack by new races of the pathogen.

As yet, race-nonspecific rust resistance is in the developmental stage in most crops, although it has played a major role against maize rusts. This latter success is the result of the natural interplay of many "minor" genes for resistance in the maize population. This success was obtained, fortuitously, by breeding maize in the field (maize grows poorly in the glasshouse), not as the result of a conscious effort to breed for such resistance. What was feared to be the start of a pandemic of southern maize rust[1] in the 1950s (see Fig. 5.5) following the introduction of *Puccinia polysora* into Africa and east Asia gradually declined in importance. This resulted in large part from selection pressure on the great variety of resistance genes in the naturally self-pollinated maize population. Those genes were selected for and stabilized resistance at a tolerable level.

Similarly, the great genetic diversity in oats in Israel has provided sustained resistance to crown rust. The magnitude and diversity of rust resistance genes, both of race-specific and race-nonspecific nature, in that center of origin of oats, have provided an invaluable contribution to plant pathologists and breeders alike. At the present time, numerous cereal cultivars are being developed that possess race-specific genes for resistance in a genetically diverse background of race-nonspecific or so-called minor genes for resistance. It is hoped that such cultivars will eventually provide the maximum protection that can be afforded by combining both genetic systems in single cultivars.

1. Three important maize rusts, each caused by different pathogens, are common maize rust (*Puccinia sorghi*), southern maize rust (*P. polysora*), and tropical maize rust (*Physopella zeae*).

REFERENCES

Buller, A. H. R. 1915. The fungus lore of the Greeks and Romans. Trans. Br. Mycol. Soc. 5:31–66.

Spencer, D. M. 1980. Parasitism of carnation rust (*Uromyces dianthi*) by *Verticillium lecanii.* Trans. Br. Mycol. Soc. 74:191–194.

Tsuneda, A., and Hiratsuka, Y. 1979. Mode of parasitism of a mycoparasite, *Cladosporium gallicola,* on western gall rust, *Endocronartium harknessii.* Can. J. Plant Pathol. 1:31–36.

FURTHER READING

General

Agrios, G. N. 1978. Plant Pathology. New York, San Francisco, London: Academic Press.

Roberts, D. A., and Boothroyd, C. W. 1972. Fundamentals of Plant Pathology. San Franciso: W. H. Freeman.
Stakman, E. C., and Harrar, J. G. 1957. Principles of Plant Pathology. New York: Ronald Press.
Tarr, S. A. J. 1972. Principles of Plant Pathology. New York: Winchester Press.
Western, J. H., ed. 1971. Diseases of Crop Plants. New York: John Wiley & Sons.
Zadoks, J. C., and Schein, R. D. 1979. Epidemiology and Plant Disease Management. New York, Oxford: Oxford Univ. Press.

Section 7.1 (Eradication)

Anderson, R. L. 1973. A summary of white pine blister rust research in the Lake States. U.S. Dep. Agric. For. Serv., Gen. Tech. Rep. NC-6.
Stakman, E. C., and Harrar, J. G. 1957. Principles of Plant Pathology. New York: Ronald Press.
Wellman, F. L. 1972. Tropical American Plant Disease. Metuchen, N.J.: Scarecrow Press.
Zadoks, J. C. 1965. Epidemiology of wheat rusts in Europe. FAO Plant Prot. Bull. 13:97–108.

Section 7.2 (Cultural Practices)

Forsberg, J. L. 1975. Diseases of Ornamental Plants. Urbana, Chicago, London: Univ. of Illinois Press.
Horner, C. E. 1965. Control of mint rust by propane gas flaming and contact herbicide. Plant Dis. Rep. 49:393–395.
Pirone, P. P. 1978. Diseases and Pests of Ornamental Plants. New York, Chichester, Brisbane, Toronto: John Wiley & Sons.
Westcott, C. 1971. Plant Disease Handbook. New York, Cincinnati, Toronto, London, Melbourne: Van Nostrand Reinhold.

Section 7.3 (Chemical Treatments)

Evans, E. 1968. Plant Diseases and Their Chemical Control. Oxford, Edinburgh: Blackwell Scientific Publications.
Forsberg, J. L. 1975. Diseases of Ornamental Plants. Urbana, Chicago, London: Univ. of Illinois Press.
Heagle, A. S. 1973. Interactions between air pollutants and plant parasites. Annu. Rev. Plant Pathol. 11:365–388.
Marsh, R. W., ed. 1977. Systemic Fungicides. London, New York: Longman.
Pirone, P. P. 1978. Diseases and Pests of Ornamental Plants. New York, Chichester, Brisbane, Toronto: John Wiley & Sons.
Westcott, C. 1971. Plant Disease Handbook. New York, Cincinnati, Toronto, London, Melbourne: Van Nostrand Reinhold.

Section 7.4 (Microbiologic Agents)

Baker, K. F., and Cook, R. J. 1974. Biological Control of Plant Pathogens. San Francisco: W. H. Freeman.

Littlefield, L. J., and Heath, M. C. 1979. Ultrastructure of Rust Fungi. Chap. 2. New York, London: Academic Press.

Section 7.5 (Plant Breeding)

Browning, J. A. 1974. Relevance of knowledge about natural ecosystems to development of pest management programs for agro-ecosystems. Proc. Am. Phytopathol. Soc. 1:191–199.

Browning, J. A.; Simons, M. D.; and Torres, D. 1978. Managing host genes: Epidemiological and genetic concepts. In Plant Diseases, vol. 1, ed. J. G. Horsfall and E. B. Cowling, Chap. 11. New York, San Francisco, London: Academic Press.

Day, P. R. 1974. Genetics of Host-Parasite Interaction. San Francisco: W. H. Freeman.

Frey, K. J.; Browning, J. A.; and Simons, M. D. 1977. Management systems for host genes to control disease loss. Ann. N.Y. Acad. Sci. 287:255–274.

International Atomic Energy Agency. 1969. Induced mutations in plants. Proc. Symp. Nature, Induction and Utilization of Mutations in Plants, Pullman, Wash.

______. 1970. Manual on mutation breeding. Tech. Rep. Ser. 119.

Nelson, R. R., ed. 1973. Breeding Plants for Disease Resistance. University Park, London: Pennsylvania State Univ. Press.

Robinson, R. A. 1976. Plant Pathosystems. Berlin, Heidelberg, New York: Springer-Verlag.

Simons, M. D. 1979. Modification of host-parasite interactions through artificial mutagenesis. Annu. Rev. Phytopathol. 17:75–96.

EPILOGUE

To quote Buller (1915), "Today the children of men still worship many strange gods, but Robigus, the stern Rust-god, is not among them. The besom [broom] of science has stretched even to high Olympus and has ruthlessly swept him into the dust-bin of oblivion." Despite this ignoble descent of Robigus, his reputed vengence is still wrought upon the world annually, evincing that man is only a part of nature, not its master. When we fully realize that fact and so design our teaching, research, and control efforts, perhaps only then will the scourges of the Uredinales as well as Robigus inhabit that same "dust-bin of oblivion."

INDEX OF RUST FUNGI

SUBJECT INDEX